AF454345

Production Technology

Production Technology

Dr. L. Krishna Reddy

M.E., Ph.D.,

Former Professor,

Department of Mechanical Engineering,

KSRM College of Engineering,

Kadapa, Andhra Pradesh.

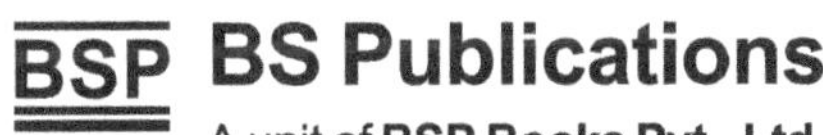

BS Publications

A unit of **BSP Books Pvt., Ltd.**

4-4-309/316, Giriraj Lane, Sultan Bazar,
Hyderabad - 500 095
Phone : 040 - 23445605, 23445688

© 2015, *by Publisher*

All rights reserved. No part of this book or parts thereof may be reproduced, stored in a retrieval system or transmitted in any language or by any means, electronic, mechanical, photocopying, recording or otherwise without the prior written permission of the publishers.

Published by :

 **BS Publications**

A unit of **BSP Books Pvt., Ltd.**

4-4-309/316, Giriraj Lane, Sultan Bazar,
Hyderabad - 500 095
Phone : 040 - 23445605, 23445688

e-mail : info@bspbooks.net

ISBN : 978-93-52300-29-7 (HB)

Dedicated to his holiness

Sri Sri Sri Ganapathi Sachchidananda Swamiji
Datta Petham
Mysore

Preface

The text book "PRODUCTION TECHNOLOGY", has been written to give a comprehensive knowledge to the students of Mechanical, Production and Automobile Engineering at Polytechnic and Engineering Colleges. This volume provides information on manufacturing processes to the budding students in a very simple way. The material is presented systematically so that the students can grasp the concepts very easily and keep in their minds for a long time. And also the scope of the book is to give the preliminary knowledge on various production processes to the learning students.

The author is extremely thankful to his colleague Dr. B. Sudheer Prem Kumar, Professor, Dept. of Mechanical Engineering, JNTU College of Engineering, Hyderabad, for editing the manuscript and for drawing line diagrams.

The author wishes to acknowledge Dr. C.S. Reddy, Principal, KSRM College of Engineering, Kadapa for his constant encouragement.

The author is also thankful to his wife Saraswathy for her inspiration, co-operation, patience and moral support in completing this text book.

All comments and suggestions for the improvement of the book will be highly appreciated.

- Author

Contents

Chapter 4

Mechanical Working of Metals 63

Chapter 5

Foundry 82

Chapter 6

Welding 168

1   FERROUS MATERIALS

Ferrous materials contain iron as the base metal. In both metallurgical and mechanical industries ferrous materials are extensively used. The various ferrous materials are classified as follows :

   (a) Pig Iron

   (b) Wrought Iron

   (c) Plain Carbon steels

   (d) Alloy steels

   (e) Cast irons

1.1 PIG IRON

Pig iron is produced by reducing iron ores such as hematite or magnetite in blast furnace. So pig iron is the first product in converting iron ores into useful metals or alloys. Pig iron obtained from blast furnace will have the following composition.

Carbon	–	3.5 to 4 per cent
Silicon	–	1 to 3 per cent
Manganese	–	0.1 to 1 per cent
Phosphorus	–	0.2 to 1.5 per cent
Sulphur	–	less than 0.5 per cent
Iron	–	Remaining

Pig iron is used as a raw material for the production of wrought iron, steel and cast iron.

1.2 WROUGHT IRON

Wrought iron is highly refined metallic iron containing a little iron silicate slag, which is distributed through out the iron in fine fibrous and film form. It's carbon content varies from 0.12 to 0.15 per cent. Wrought iron is produced from pig iron either by puddling process or by Byers process.

Wrought iron possess high ductility and it can be easily forged and welded.

Wrought iron is mostly used for making small size water pipes and fittings, corrugated sheets and ornamental sheets and also used for making chains, bolts, nails, anchors etc.

1.3 PLAIN CARBON STEELS

Steel is an alloy of iron and carbon with small amounts of silicon, manganese, phosphorus and sulphur.

A plain carbon steel is a steel in which the properties mainly depend upon the amount of carbon it contains and no other alloying elements are added. The plain carbon steels are classified on the basis of their carbon content as :

   (a) Low carbon steels or mild steels   – 0.08 to below 0.3 per cent carbon.

   (b) Medium carbon steels   – 0.3 to below 0.6 per cent carbon

   (c) High carbon steels   – 0.6 per cent carbon and above, usually upto 1.5 per cent carbon.

1.3.1 Low Carbon Steels or Mild Steels

The plain carbon steels which contain carbon in the range of 0.08 to below 0.3 per cent are called as low carbon steels or mild steels.

Mild steels are not much affected by heat treatment processes. Usually mild steels possess high ductility, good machinability and weldability. The typical mechanical properties of mild steels are as follows:

    Ultimate tensile strength   –   4000 - 6000 kg/cm^2

    Brinnell hardness number   –   150

Mild steels are used for making wires, rivets, screws, bolts, nuts, sheets, plates, tubes, rods and for general work shop purposes.

1.3.2 Medium Carbon Steels

The plain carbon steels which contain carbon in the range of 0.3 to below 0.6 per cent are called as medium carbon steels.

The mechanical properties of medium carbon steels can be further improved by proper heat treatment process. Medium carbon steels are stronger than mild steels but possess less ductility and machinability than mild steels. These steels can be easily forged and welded. The typical mechanical properties of a 0.5 per cent carbon steel are as follows:

Ultimate tensile strength	–	7500 kg/cm^2
Brinnell hardness number	–	200

Medium carbon steels are used for making nuts, bolts, springs, agricultural tools, large forging dies etc.

1.3.3 High Carbon Steels

The plain carbon steels which contain carbon in the range of 0.6 to 1.5 per cent are called as high carbon steels.

The mechanical properties of high carbon steels can be further improved by proper heat treatment process. Because of high carbon content, high carbon steels possess high strength and hardness but they have less ductility and machinability. The typical mechanical properties of a 0.8 per cent carbon steel are as follows:

Ultimate tensile strength	–	$9500 – 11000 \text{ kg/cm}^2$
Brinnell hardness number	–	550

High carbon steels are used for making springs, metal cutting tools, press work dies and punches, forging dies, hand files, drills etc.

1.4 ALLOY STEELS

Alloy steels are the types of steels in which elements other than the carbon and iron are present in sufficient quantities to improve the properties of steels. The utility of alloy steels lies in the fact that they permit a much wider range of physical and mechanical properties than is possible in plain carbon steels. In plain carbon steels increase of tensile strength and hardness by an increased carbon content or heat treatment is liable to be accompanied by a relative loss of toughness and ductility. But in case of alloy steels these losses are not there and some alloy steels can even have tensile strength of 31.14 tons/sq.cm which is not possible with plain carbon steels.

Alloying elements are added to steels for many purposes.

The most important purposes are :

(a) To increase hardenability

(b) To improve mechanical properties

(c) To increase corrosion and wear resistance

(d) To improve magnetic properties.

1.4.1 Influence of Alloying Elements in Steel

The influence of various alloying elements in steel is as follows:

(i) **Aluminium :** Aluminium acts as a deoxidiser and restricts grain growth. When heated in contact with nitrogen, aluminium forms hard aluminium nitrides. When aluminium is added in small amounts, it increases strength but too much aluminium results in embrittlement. From 2 to 5%, it imparts resistance to heat and oxidation.

(ii) **Boron :** Boron is a very powerful hardening agent but is very expensive. Hence to be economical, boron is used in minute amounts not over 0.003%. When too much boron is added it produces brittleness and hot shortness.

(iii) **Chromium :** Chromium forms hard carbides and imparts good wear and corrosion resistance and high hardness. When added in small amounts, it increases strength, impact resistance and toughness. Chromium decreases machinability and decreases hardening temperature range unless balanced with nickel. When chromium is added in amounts in excess of 5 per cent, the high temperature properties and corrosion resistance of the steel are greatly improved.

(iv) **Cobalt :** Cobalt produces red hardness by retaining hard carbides at high temperatures. But it has a tendency to decarburise steel during heat treatment. It increases hardness and strength but too much of it decreases impact resistance of the steel. Cobalt increases residual magnetism in steel.

(v) **Copper :** Addition of copper in steel varies from 0.1 to 0.4 per cent. Copper increases resistance to atmospheric corrosion and also acts as a strengthening agent.

(vi) **Manganese :** Manganese, which is one of the least expensive alloying elements, is widely used in steel production both for deoxidation and desulphurisation of molten steel. It is generally added as ferromanganese (80 per cent Mn, 6 per cent C, balance Fe) or as spiegeleisen (5-20 per cent Mn, 5 per cent C, balance Fe). Manganese is present between 0.40 to 1.00 per cent in most of the commercial carbon steels. When manganese content exceeds 1.00 per cent then it is regarded as a deliberate alloying element. The main function of manganese is to minimise the harmful effects of sulphur. Manganese combines with sulphur and forms MnS which is less harmful than FeS. The presence of MnS globules improves the machinability of steel and also wear resistance.

Addition of manganese increases tensile strength and hardness. If manganese is present in large amount, it lowers the critical temperature and when this steel is quenched rapidly it becomes austenitic and possess toughness with ductility.

(vii) **Molybdenum :** In low nickel, low chromium steels, addition of molybdenum reduces the tendency to temper brittleness. The presence of molybdenum in nickel-chromium steels, reduces the transformation rates further and therefore contributes considerably to depth of hardening.

Molybdenum dissolves in ferrite which it strengthens considerably and also forms hard carbide Mo_2C as well as double carbides such as Fe_4Mo_2C and $Fe_2Mo_2C_6$. Molybdenum increases the high temperature strength and impact resistance at high temperatures. It enhances the corrosion resistance of stainless steels particularly to chloride solutions. Molybdenum also increases machinability of carbon steels. The amount of molybdenum usually employed in any of the molybdenum steel varies between 0.15 and 0.60 per cent.

(viii) **Nickel :** Nickel is widely used in alloy steels in quantities upto about 5.0% to increase strength and toughness. Nickel lowers the critical temperatures of steel, retards decomposition of austenite and does not form any carbides. The presence of nickel in large amounts increases resistance to oxidation at high temperatures. Nickel also decreases the machinability.

(ix) **Phosphorus :** Phosphorus is always present in varying amounts in all steels. It dissolves in ferrite and increases its tensile strength and hardness. Phosphorus separates as iron phosphide (Fe_3P) when added in excess of the solubility limit. Iron phosphide is a hard and brittle phase and therefore addition of phosphorus in large amount increases brittleness and cold shortness of steel. Due to this, phosphorus is kept below 0.05 per cent in most of the steels.

Besides increasing the tensile strength and hardness, phosphorus also improves machinability and resistance towards atmospheric corrosion.

(x) **Silicon :** Silicon is present in almost all the steels in varying amounts as a cheap deoxidiser and graphitiser. Silicon dissolves in ferrite and is not a carbide former. Iron-silicon (0.5 to 4.5%) alloys have a high magnetic permeability but a very low hysteresis and hence they are widely used in the electrical industry as magnetically soft materials for transformer and generator laminations.

Silicon forms hard iron silicides and when present in large amounts it gives high hardness, wear resistance and acid resistance but causes brittleness.

(xi) **Sulphur :** Sulphur is a harmful element in steel and hence it is restricted to 0.05 ercent in most of the steels. The main harmful effect of sulphur is, it induces hot shortness in steel i.e., if excess sulphur is present, cracks may form during hot rolling and forging operations. The only advantage of sulphur is that it improves the machinability of steels.

(xii) **Tantalum :** Tantalum is added in some special steels to improve resistance to scaling at high temperatures.

(xiii) **Titanium :** Titanium is a strong carbide former, effectively inhibits grain coarsening and acts as a grain refiner. In order to prevent the precipitation of chromium carbides, titanium is added to stainless steels.

(xiv)   **Tungsten :** Tungsten is a strong carbide former and forms extremely hard and very stable carbides W_2C, WC and $Fe_4 W_2C$. These carbides imparts wear and abrasive resistance to the steel. Tungsten inhibits grain growth and therefore has a grain refining effect.

Tungsten retards the softening of martensite during tempering and therefore it gives red hardness. Due to this reason tungsten is an important constituent of most of the high speed tool steels and hot working die steels in which it develops red (high temperature) hardness following suitable heat treatment.

(xv)   **Vanadium :** Vanadium is a powerful deoxidizer, a strong carbide former, inhibits grain growth and very expensive. Additions of about 0.05 per cent vanadium produces fine grain structure by retarding grain growth.

Vanadium forms hard carbide VC which imparts excellent wear resistance and resistance to tempering.

1.4.2 Special Purpose Steels

During service most of the commercial metals and alloys are subjected to abnormal environments such as the actions of concentrated acids, alkalies and corrosive gases at elevated temperatures, prolonged exposure at temperatures above red heat, severe cold working conditions within the range of plastic deformations etc.

Special purpose steels which are developed to meet these conditions are :

(a) High speed tool steel

(b) High chromium steel

(c) Stainless steels

(d) High nickel steels and special nickel alloys

(e) Austenitic high manganese steels (or) Hadfield steels

(f) Heat resistance steels

High Speed Tool Steels

High speed tool steels are developed to withstand the heat produced when a cutting tool is used at a high speed. Normally a large amount of heat is produced due to friction between tool and the work piece. Only high speed tool steels can withstand this heat without losing hardness. They cut at twenty times the speed of a plain carbon tool without losing their sharpness.

The conventional brand of high speed tool steel will have the following composition:

Carbon   –   0.70 per cent

Tungsten   –   18 per cent

Chromium – 4 per cent

Vanadium – 1 per cent

Tungsten provides toughness, wear resistance and cutting ability, chromium serves to increase the hardenability and vanadium is for grain refinement. This variety of high speed tool steel can be used for machining operations on steel and non ferrous materials. Twist drills, taps, reamers, milling cutters and similar tools are manufactured out of this brand.

In order to impart additional red hardness thereby permitting the tool to maintain a higher hardness and increased cutting ability at elevated temperatures, 6 to 10 per cent cobalt is added to the above composition.

Now a days tungsten is replaced by molybdenum to produce a high speed tool steel at much lower cost. The molybdenum high speed steels contain 1.5 to 6 per cent tungsten, 4 to 9 per cent molybdenum and between 3.5 and 4.5 per cent chromium. From the stand point of fabrication and tool performance there is little difference between the molybdenum and tungsten grades. The important properties of red hardness, wear resistance and toughness are about the same. The molybdenum steels are lower in price than tungsten steels and over 80 per cent of all the high speed steel produced is of the molybdenum type.

High Chromium Steels

The addition of chromium to the plain carbon steels improves hardenability, strength and wear resistance. Chromium has a marked resistance towards corrosion and heat and has good mechanical properties at elevated temperatures.

High alloy chromium steels are broadly classified into three types:

(A) The first variety are those containing upto 10 per cent chromium and upto 4 per cent of one or more of the elements like silicon, nickel, molybdenum and tungsten. Normally in this type of steels, the total percentage of chromium and other alloying elements is not less than 5 per cent.

In oil refineries, low carbon steel with about 5 per cent chromium, 0.5 per cent silicon and 0.6 per cent molybdenum is widely used.

The most important commercial steel of this type is **Silchrome**. Silchrome is usually used for the manufacture of exhaust valves of internal combustion engines and it contains about 8 to 10 per cent chromium and 1.5 to 4.0 per cent silicon. Silchrome has good strength and toughness at slightly elevated temperatures and resists the corrosive action of the exhaust gases.

(B) The second variety are those containing about 12 to 18 per cent chromium. Examples are cutlery steels and surgical steels. The common cutlery steels contain 14 to 16 per cent chromium and about 0.25 to 0.50 per cent carbon. Surgical steels have the same composition, but they contain in addition about one per cent nickel. These steels are normally oil quenched from 1000 °C and tempered at about 450 °C to get a hardncess of about 480 Brinell.

(C) The third variety are high heat resisting steels containing about 20 to 30 per cent chromium and about 0.40 per cent carbon. For making boxes for annealing and heat treatment operations, these steels are usually employed.

Stainless Steels

These alloy steels which are highly resistant to corrosion and oxidation are known as stainless steels. In addition, they often have good creep strength.

There are several fairly distinct types of stainless steels having large percentages of chromium (more than 10 per cent) and some times of nickel (0 to 26 per cent). But according to their microstructure, stainless steels are broadly grouped into three types.

(A) Ferritic stainless steels

(B) Martensitic stainless steels

(C) Austenitic stainless steels

It is observed that for %C > 12.7 per cent, the ferrite phase becomes stable over the entire temperature range upto the melting point. But in the presence of carbon, to produce an all ferrite microstructure, it is required that %C – 17 x %C should be greater than 12.7 per cent. If %C – 17 x %C is less than 12.7 per cent, then we obtain martensitic stainless steels because martensite forms from the high temperature austenite during cooling to room temperature. To obtain austenitic stainless steels, the austenite stabilizers such as Ni and Mn are added in sufficient quantities to make austenite stable at room temperature.

(A)   Ferritic Stainless Steels

The stainless steel is in the ferrite state when %C – 17 x %C is greater than 12.7 per cent. These steels cannot be heat treated, since austenite does not form at any temperature and can only be hardened by cold working. Grain refinement is done by recrystallization following cold work.

Typical examples of ferritic stainless steels are 16% Cr, 0.12 %C, and 25% Cr, 0.2 %C. Due to high chromium content these steels exhibits excellent corrosion and oxidation resistance. Typical mechanical properties of 16% Cr, 0.12 % C ferritic stainless steel in the annealed condition are:

Yield stress	–	350 Mpa
Tensile stress	–	550 Mpa
% Elongation	–	30

These steels are low cost (since expensive nickel is not present), soft, ductile, malleable and magnetic in character.

Ferritic stainless steels are used for house hold articles, dairy machinery, aeroplane and automobile fittings, stainless nuts and bolts, chemical plants to resist the action of nitric acid etc., and furnace parts which are not subjected to high stresses.

(B) Martensitic Stainless Steels

For martensitic stainless steels %C – 17 x %C should be less than 12.7 per cent and these steels respond to heat treatment. They contain chromium from 12 to 18 per cent and carban from 0.15 to 1.2 per cent. The typical as quenched properties of 12 to 14 per cent chromium, 0.15 per cent carbon martensitic steel are :

Yield stress	–	1200 Mpa
Tensile stress	–	1300 Mpa
% Elongation	–	5

Martensitic stainless steels are hard, wear resistant and magnetic in character. These are used for springs, ball bearings, valves, razers and razer blades, surgical instruments, cutting tools, cutlery items etc.

(C) Austenitic Stainless Steels

Austenitic stainless steels usually contain nickel and some times this nickel is partly or fully replaced by the cheaper manganese. Since nickel and manganese are austenite stabilisers, these steels are austenitic at room temperature. Hence these steels are called as austenitic stainless steels.

The most commonly used type is 18-8 austenitic stainless steel which contain 18 per cent chromium, 8 per cent nickel and 0.08 per cent carbon. 18-8 austenitic stainless steels have excellent forming characteristics due to high strain hardening rate and the uniform elongation is also large.

Typical mechanical properties of 18-8 austenitic stainless steel are :

Yield stress	–	240 Mpa
Tensile stress	–	600 Mpa
% Elongation	–	65

This steel is used in a wide variety of applications in the chemical industry and for house hold and sanitary fittings.

The corrosion resistance of the austenitic stainless steels is usually better than that of the martensitic or ferritic stainless steels.

High Nickel Steels and Special Nickel Alloys

The most widely used high nickel steels are:

(a) Elinvar

(b) Perm alloy

(c) Platinite

(d) Hypernik

These four alloy steels are discussed below:

(a) Elinvar : Elinvar is an alloy steel which contains 32 per cent nickel, 5 per cent chromium and 2 per cent tungsten so that the effect of temperature variations due to climatic changes on the modulus of elasticity is zero. Elinvar possess excellent corrosion resistance and is used for the manufacture of hair springs of watches and many other sensitive instruments where temperature variations should not affect the accuracy.

(b) Perm Alloy : Perm alloy contains 80 per cent nickel and possess magnetic permeability 30 times that of soft iron. Hence this alloy responds even under weak magnetizing forces. Because of this property, this alloy is used extensively in communication engineering especially in long distance telephony.

(c) Platinite : Platinite contains 45 per cent nickel and 0.15 per cent carbon. This alloy has exactly the same coefficient of expansion as glass and is widely used in the manufacture of armoured glass.

(d) Hypernik : Hypernik contains 55 per cent nickel and possess high magnetic permeability and finds wide application in radio transformers.

Austenitic High Manganese Steels (or) Hadfield Steels

These steels, manufactured by electric arc process, usually contain 1.0 to 1.3 per cent carbon and 11 to 14 per cent manganese. These are also called as Hadfield steels named after the inventor Sir Robert Hadfield (1882).

Manganese stabilizes the austenite phase by depressing the A3 transformation temperature. This results in austenitic structure in Hadfield steels. Manganese also forms manganese carbide similar to iron carbide. Therefore the structure of Hadfield steels which are allowed to cool slowly to room temperature, consists mostly of austenite and manganese carbides. Hence to prevent the precipitation of carbides and to keep the carbon in solution, Hadfield steels are water quenched from 1050°C. The quenched steel will have the following properties.

Tensile strength	–	94 to 118 kg/sq.mm
% Elongation	–	60 to 70
Brinnell hardness number	–	190 to 225

Being austenitic, Hadfield steel is extremely tough and shock resistant. The Brinnell hardness of this steel increases rapidly from 200 to as high as 550 Brinnell during cold working.

Hadfield steel is available as castings, forgings or hot rolled sections. It is difficult to machine this steel because of its tendency to harden as soon as machining is attempted.

Hadfield steel find wide use in rock crushing machinery, dredging equipment etc.

Heat Resistance Steels

Heat resisting steels are required for a wide variety of applications such as pipes in steam power plant, aeroplane and automobile valves, furnace conveyors, retorts, gas turbines and glass making machinery. Heat resisting steels should possess the following properties:

(i)   Good creep resistance

(ii)   Resistance to oxidation and scaling

(iii)   Specific properties relating to particular application e.g. machinability, weldability, fatigue properties and coefficient of thermal expansion.

In order to meet these requirements a number of heat resisting steels have been developed which may be classified as follows:

(a) Low alloy steels (0.5% Molybdenum)

(b) Chromium-silicon valve steels

(c) Plain chromium steels (12 - 30% chromium)

(a) **Low Alloy Steels :** The main applications of these steels are for pipes in steam plants where service temperatures are in the range of 400 °C to 550 °C. Typical compositions of suitable heat resisting steels of this type are those containing (a) 0.5% Mo (b) 0.5% Mo + 1% C (c) 0.5% Mo + 0.25% V, (d) 0.5% Mo + 0.25% C. These additions are all carbide forming elements.

(b) **Valve Steels :** Chromium-silicon steels such as **silchrome** (0.4% C, 8% Cr, 3.5% Si) and **valmax** (0.5% C, 8% Cr, 3.5% Si, 0.5% Mo) are used for automobile valves. They possess good resistance to scaling at a dull red heat, although their strength at elevated temperatures is relatively low.

For aero engines and marine diesel engine valves 13% nickel, 13% chromium, 3% tungsten valve steel is usually used.

(c) **Plain Chromium Steels :** These consist of the martensitic chromium steels with 12 –13% Cr and the ferritic chromium steels with 18 – 30% C.

Plain chromium steels are very good for oxidation resistance at high temperatures as compared to their strength which is not high under such conditions. The maximum operating temperature for the martensitic steels is about 750°C, whereas for ferritic steels it is about 1000 – 1150 °C. Such steels have a good resistance to sulphurous atmospheres.

1.5 CAST IRONS

Cast iron is an alloy of iron and carbon. Carbon content in cast iron is above 2 per cent and if carbon content is below 2 per cent it is called as steel. Carbon may occur in cast irons as cementite (iron carbide) which is called as combined carbon or carbon may also occur in cast irons as graphite which is called as free carbon.

Cast iron is one of the most important engineering material because it offers a wide range of properties.

The following are the definite advantages and characteristics which secured a unique place to cast irons in engineering industry:

(a) Cast irons are cheap.

(b) Cast irons offers wide range of mechanical properties such as high strength, high hardness, good machinability, good wear resistance, good abrasion resistance and good corrosion resistance.

(c) Cast irons are highly desirable for casting purposes because of their foundry properties such as high casting yield, good fluidity, low shrinkage, casting soundness and ease of production.

(d) Cast irons have lower melting temperature (1140 °C to 1250 °C).

(e) Cast irons have high damping capacity which absorb vibrational stresses.

1.5.1 Cast Irons Microstructure

All the properties of cast irons depend upon their microstructure and these properties can be varied by changing the micro-structure.

Based on the phases that are present in the microstructure, the cast irons are differentiated into various types such as gray, white, malleable, spheroidal etc.

The most important phases of cast irons microstructure are discussed below:

Graphite

When carbon occurs in cast irons as free or elemental form then it is termed as graphite. This graphite occurs in several forms. In gray cast iron, graphite appears as flakes, in malleable iron, temper carbon or graphite aggregates are present and in spheroidal graphite cast iron, graphite appears as spheroids. The amount, size, shape and distribution of the graphite in cast irons influence their properties considerably. For example, flake graphite reduces ductility whereas temper carbon or spheroidal graphite does not reduce ductility to the extent of flake graphite.

Cementite

When carbon occurs in cast irons as chemically combined form Fe_3C, then it is termed as cementite. This cementite occurs in white or chilled cast irons. Since cementite phase is very hard and brittle, the resultant white or chilled cast irons become very hard and brittle.

Ferrite

Solid solution of carbon in α iron is called as ferrite. Ferrite phase is relatively soft, ductile and of moderate strength.

Ferrite phase in cast irons may occur as free ferrite or as ferrite in pearlite phase. Free ferrite occurs in malleable iron and spheroidal iron. Ferrite occurs as a constituent of pearlite phase in gray cast irons.

Pearlite

A mixture of ferrite and cementite is called as pearlite, and these ferrite and cementite appear in microstructure of cast irons as alternate layers. Pearlite in cast irons has good strength, moderately hard and possess some ductility.

Austenite

Solid solution of carbon in γ iron is called as austenite. Austenite is not a stable phase at room temperature and under slow cooling it changes to pearlite. Hence austenite, as a part of microstructure at room temperature, can be obtained in cast irons which are specially alloyed with nickel since nickel is an austenite stabilizer and makes austenite stable at room temperature.

Steadite

Presence of phosphorus causes the formation of steadite phase in the microstructure of cast irons especially in gray cast iron. Phosphorus present in cast irons combine with iron and forms iron phosphide (Fe_3P).

An eutectic of iron and iron phosphide is called as steadite. Steadite has a low melting point (980 $^\circ$C) and is the last to solidify and therefore occupies interdendritic regions.

Steadite, like cementite is very hard and brittle and therefore it reduces toughness and induces brittleness in cast irons. Hence excessive phosphorus content increases the hardness and brittleness of gray iron due to the formation of steadite.

1.5.2 Alloying Elements and their Effect

Alloying elements added to cast irons may be divided into two groups:

(a) Graphite forming elements which promote the formation of graphite (graphitisation) such as carbon, silicon, nickel, copper etc.

(b) Carbide forming elements which promote the formation of cementite such as sulphur, chromium, vanadium etc.

(a) Graphite forming elements

(i) Carbon

In cast irons carbon is present either in the combined form (Fe_3C) or as free graphite. Iron carbide (Fe_3C) induces strength and hardness whereas graphite induces softness.

When carbon content is more, the tendency of graphitisation is more which results in the formation of gray cast iron.

And when carbon content is less, the tendency to form cementite (Fe_3C) is more which results in the formation of white cast iron.

(ii) Silicon

Silicon is a strong graphitising element and hence it tends to break iron carbide (Fe_3C) into ferrite and graphite.

$$Fe_3C \rightarrow 3Fe + C \text{ (graphite)}$$

The amount of silicon controls the amount of graphitization. Normally amount of silicon varies from 0.5 to 3.0 per cent depending upon the amount of graphitization required. Lower the amount of silicon, the cast iron solidifies as white cast iron whereas higher the amount of silicon, the cast iron solidifies as gray cast iron.

Elements like manganese, chromium, molybdenum, vanadium etc., reduce the graphitizing effect of silicon and are called chill inducing elements. Chilling normally produces white cast iron.

Elements like copper, nickel, titanium, aluminium, zirconium etc., increase the graphitizing effect of silicon and are called chill restraining elements. Restraining of chilling normally produces gray cast iron.

Silicon improves tensile strength and hardness. Highest tensile strength is obtained with a carbon content of 2.75 per cent and a silicon content of 1.5 per cent.

(iii) Nickel

Nickel is a chill restraining or graphitizing element. It promotes graphitization.

Addition of 1 to 2 per cent nickel to cast iron, eliminates hard spots, increases fluidity, improves machinability and heat resistance.

Nickel gives good surface finish to cast iron castings.

(iv) Copper

Copper is also a chill restraining or graphitizing element. It promotes graphitization.

Copper imparts corrosion resistance to cast iron. Due to slight solubility in cast iron, copper is added in small quantities.

(v) Manganese

From pig iron, manganese enters into cast iron. And sometimes additions of ferro manganese may be made to obtain the desired proportion of manganese.

The main function of manganese is to neutralize the effect of sulphur by combining with sulphur and forming manganese sulphide. Sulphur is a carbide forming element or hardner and hence the addition of manganese causes softening of the metal due to neutralization of sulphur effect.

(b) Carbide forming elements

(i) Sulphur

Sulphur is a powerful chill inducing or carbide forming element. It promotes the formation of cementite (Fe_3C) in cast iron.

Sulphur is usually present in cast iron in quantities less than 0.12 per cent.

(ii) Chromium

Chromium is a chill inducing or carbide forming element. It is helpful to stabilize iron carbide in white cast iron.

When high resistance to heat and corrosion is required, chromium is added in appreciable quantities to cast iron.

(iii) Vanadium

Like chromium, vanadium is also a chill inducing or carbide forming element. It strongly stabilizes iron carbide at high temperatures. It improves hardness, tensile strength, and wear resistance but decreases machinability.

1.5.3 Family of Cast Irons or Types of Cast Irons

Cast irons are classified based on the form of carbon in the micro structure. The various types of cast irons are as follows:

(a) Gray cast iron

(b) White cast iron

(c) Malleable cast iron

(d) Spheroidal graphite cast iron (S.G. Iron)

(e) Alloy cast iron

Gray Cast Iron

After solidification, the microstructure of gray cast iron consists of a large part or all of the carbon is in the form of flakes of graphite (Fig.1.1). When gray cast iron is fractured, the fractured surfaces have gray in appearance and hence it is called as gray cast iron.

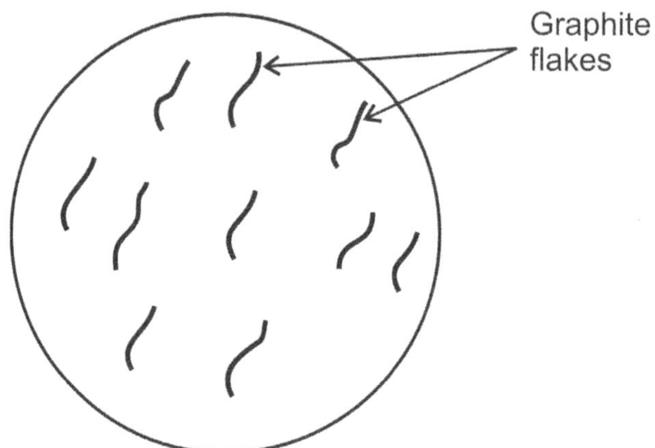

Fig. 1.1 *Micro structure of gray cast iron.*

Gray cast iron has high machinability, high damping capacity and high compressive strength but it possess low tensile strength and impact strength. Gray cast iron is used for machine tool beds and structural members loaded in compression.

White Cast Iron

After solidification, the micro structure of white cast iron consists of practically all the carbon in the chemically combined state as cementite (Fe_3C). When white cast iron is fractured, the fractured surfaces have dull white (bright) appearance and hence it is called as white cast iron (Fig. 1.2).

Fig. 1.2 *Micro structure of white cast iron.*

White cast iron is very brittle, wear resistant and have high hardness. The machinability is very low. White cast iron is usually used for parts requiring high abrasion resistant.

Malleable Cast Iron

Malleable cast iron is obtained by subjecting white cast iron to a special annealing process called malleablisation. After malleablisation, the microstructure of malleable cast iron consists of graphite in the form of small rounded nodules called temper carbon (Fig. 1.3).

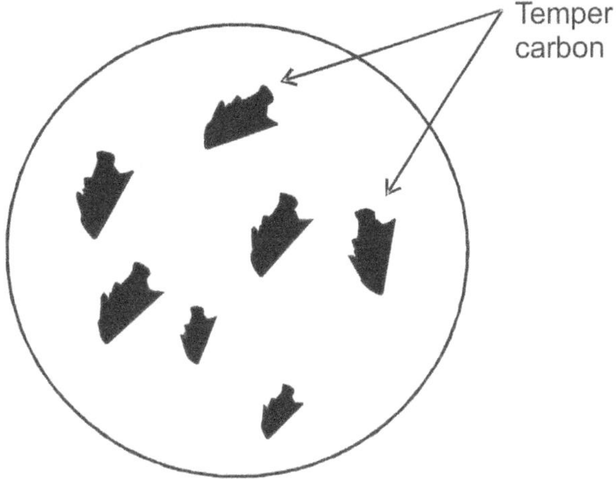

Fig. 1.3 *Micro structure of malleable cast iron.*

Because of this micro structure, malleable cast iron is stronger than gray cast iron and more resistant to bending and twisting.

Malleable cast iron is used for automobile parts, agricultural implements such as tractor and plough parts, gear housing etc.

Spheroidal Graphite Cast Iron (S. G. Iron)

S.G. iron is also called as ductile cast iron. S. G. iron is obtained by adding a small amount of magnesium or cerium to the molten gray cast iron. Magnesium or cerium causes the graphite to take a strictly rounded spheroidal form instead of flakes (Fig. 1.4) .

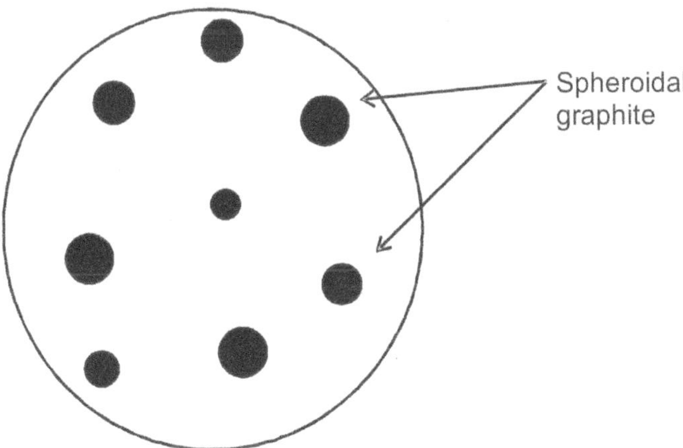

Fig. 1.4 *Micro structure of S.G. Iron.*

S.G. iron possess very high fluidity, castability, strength, toughness, ductility, wear resistance, machinability and weldability.

S.G. iron is used to produce both intricate casting as well as large castings. It is mainly used for the manufacture of heavy duty machinery, gears, dies etc.

1.5.4 Production of Spheroidal Graphite Cast Iron

Spheroidal graphite cast iron combines the principle advantages of gray cast iron such as good fluidity, low melting point, good castablity, machinability and wear resistance with the engineering advantages of steel such as high strength, ductility, toughness, hardenability and hot workability.

In the year 1948, S.G. iron was introduced to the foundry industry by the British Cast Iron Research Association (BCIRA) and the International Nickel Company (INCO). The BCIRA process used cerium and the INCO process used magnesium to produce graphite spheroids instead of flakes. After the introduction of these two processes to the foundry industry, the INCO process using magnesium is widely accepted by the world wide foundry industry and is now almost universally employed for the production of S.G. iron.

Function of Magnesium : When magnesium is added to gray cast iron, it first acts as deoxidiser and desulphuriser of the molten metal. Therefore oxygen and sulphur content of the molten metal should be kept low otherwise much of the added magnesium will be consumed by forming magnesium oxides and magnesium sulphides. And then finally magnesium produces graphite spheroids by preventing the formation of graphite flakes. Normally 0.05 per cent residual magnesium is sufficient to produce graphite spheroids.

Melting Furnace : Cupola furnace is usually used for melting gray cast iron to produce S.G. iron. Both acid cupola and basic cupola can be used.

Though acid cupola is much cheaper but the problem is in the control of sulphur. Due to acid refractory lining in the acid cupola, sulphur content can not be reduced by the acid slags which are produced in this furnace. Basic slags produced by adding lime can reduce the sulphur content, but basic slags attacks the acid lining and causes damage to the acid cupola furnace. Hence if acid cupolas are employed then external desulphurisation has to be carried out in the ladle by adding soda ash, calcium carbide or lime, to molten cast iron before subjecting to magnesium treatment. Otherwise much of the costly magnesium alloy will be consumed for desulphurisation before graphite spheroidization can occur. Acid cupola with external desulphurisation will be much cheaper than basic cupola melting since acid refractories are four to five times cheaper than basic refractories.

Though costly, basic cupola melting has a definite advantage of sulphur control. By reducing the refractory cost with the latest developments such as water cooled cupolas and hot blast, basic cupolas are also widely used for producing S.G. iron.

Addition of Magnesium : Amount of magnesium added to the base iron depends on sulphur and oxygen content. After desulphurisation and deoxidization, the retained magnesium content of around 0.05 per cent is considered to be sufficient to produce graphite spheroids.

And also, since the boiling point of magnesium is low compared to the base iron melting temperature (2800°F), the added magnesium gets vapourized on contact with the base iron and the reaction may be quite violent. Hence to prevent the loss of magnesium, magnesium is added in the form of alloys such as magnesium-nickel alloys, magnesium-silicon alloys and magnesium-ferro silicon alloys. With these alloys magnesium recovery is found to be higher.

Magnesium recovery is also dependent on the way magnesium is added to the molten metal. The usual methods employed for adding magnesium are :

(a) the open ladle method,

(b) the plunging method and

(c) the mechanical feeder method.

(a) The Open Ladle Method : The ladle used in this case is specially designed to be deep and narrow having a height to diameter ratio of 2:1. In this method, first magnesium alloy is placed in the bottom of the ladle and then molten metal is tapped on to it. In this way, the magnesium vapour can penetrate a greater depth into the molten iron.

This method is further improved and called as sand wich method. In sand wich method, the magnesium alloy is placed into a recession in the refractory bottom of the ladle. On this alloy, a steel plate, iron chips, steel punchings etc., are placed and then the molten iron is tapped into the ladle. By this way the reaction between magnesium and molten metal is delayed until the ladle is filled partially, thus magnesium recovery is improved further.

(b) The Plunging Method : In this method the magnesium alloy is placed in a container called refractory bell which is fastened to a refractory coated plunging rod. The plunging rod with magnesium alloy in the bell is plunged into the molten metal contained in a ladle.

(c) The Mechanical Feeder Method : In this method, the magnesium alloy is continuously added mechanically to the molten stream of molten metal which is emitting from cupola. This method claims greater magnesium recoveries.

Applications : S.G. Iron is used for the manufacture of heavy duty machinery, gear, dies, rolls for wear resistance and strength, track rollers, crank shafts for navy ship board etc.

1.5.5 Production of Malleable Iron

Malleable iron is a most suitable material for a design engineer due to its unique properties such as good machinability, ductility, toughness, strength and corrosion resistance.

The micro structure of malleable iron consists of temper carbon (nodular form of graphite) in a matrix of ferrite. By subjecting white cast iron to a special heat treatment process, this microstructure of malleable iron can be obtained. This heat treatment breaks the hard carbide structure to produce nodular form of graphite or temper carbon [$Fe_3C \rightarrow 3Fe + C$ (graphite)].

Malleableization

Malleable iron is produced by the heat treatment of white cast iron. This heat treatment process is called as malleableization since it converts the hard, brittle white cast iron into malleable iron.

The white cast iron microstructure at the time of heat treatment will consist of pearlite, hard carbides and some eutectic areas.

The heat treatment cycle is given in the Fig. 1.5.

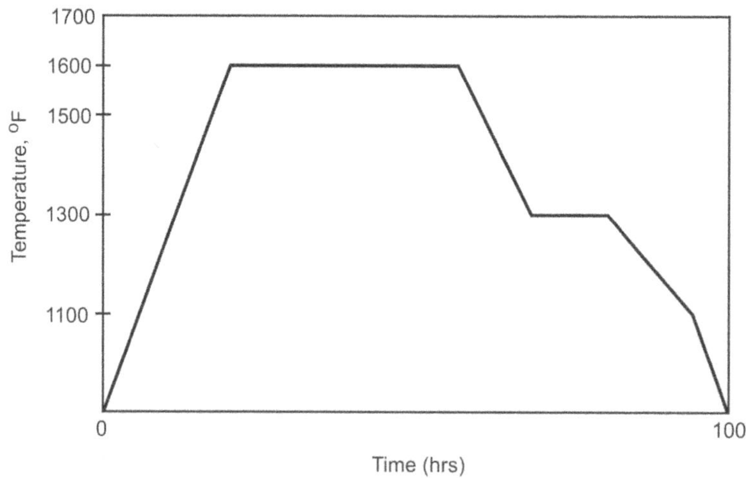

Fig. 1.5 *Heat treatment cycle.*

The heat treatment process involves three steps :

Ist step : During the first step nucleation of graphite will occur. This is a very important step since the success of heat treatment depends on graphite nuclei. Nucleation of graphite will occur during heating to the holding temperature and very early during the holding period.

IInd step : The second step is called as first stage graphitization (FSG) which occurs during holding period at a temperature of 1600 to 1700 °F. During this step the hard, massive carbides will be eliminated and growth of the graphite nuclei (which is formed in the first step) into nodular form (temper carbon) will occur at the expense of the carbides. During this step the carbide structure will disappear completely.

IIIrd step : The third step is called as second stage graphitization (SSG). This step involves slow cooling from the holding temperature to the room temperature. During this step ferritic matrix free from pearlite and carbides will form.

Thus with this heat treatment cycle malleable iron having temper carbon (nodular form of graphite) in a ferrite matrix is produced from white cast iron.

Application : Malleable iron is used in the automotive and truck industries, construction machinery and agricultural equipment producers.

1.5.6 Inoculated Cast Irons

Ferro-silicon is termed as a graphitiser or an inoculant. Inoculated cast irons are produced by adding, just before casting, a suitable amount of an inoculant like ferro-silicon. Addition of inoculant produces graphite flakes in fine form instead of coarse form as in normal gray cast iron. And also causes the graphite flakes to distribute randomly throughout the micro structure. This makes the inoculated cast irons to become more stronger than the ordinary gray cast iron. Inoculated cast irons with fine graphite flakes will have tensile strength more than 3 tonnes/cm^2.

The other inoculants usually added are calcium silicide, nickel shot, zirconium-silicon.

1.5.7 Alloy Cast Irons

To improve corrosion resistance, wear resistance and impact resistance, certain alloying elements such as nickel, chromium, silicon, molybdenum, copper, vanadium are added to cast irons.

Some of the important alloy cast irons are :

(a) Ni– resist,

(b) Ni– hard and

(c) Silal.

Ni– resist : Ni-resist is a gray iron with about 15 per cent nickel. Since nickel is an austenite stabilizer, the matrix of this cast iron becomes austenitic. This Ni– resist alloy cast iron is used for corrosion resistant applications in chemical plants.

Ni– hard : To improve abrasion resistance, Ni– hard cast iron is produced by adding nickel and chromium to the white cast iron. Addition of nickel and chromium produces a martensitic matrix in Ni – hard cast irons.

Normally Ni – hard cast irons contains about 3 to 5 per cent nickel and 1 to 3 per cent chromium. The hardness of Ni – hard cast irons will vary between 400 – 680 BHN.

Silal : To improve corrosion and oxidation resistance, about 5 per cent silicon is added to low carbon cast iron and the resultant cast iron is called as silal. Silal microstructure consists of ferrite and fine graphite.

The draw back of silal is brittleness and is overcome by adding nickel and chromium. Addition of nickel and chromium reduces brittleness and such cast irons are called Nicro silal.

Silal and nicro silal are used for carrying corrosive fluids.

1.6 MODEL QUESTIONS

1. Distinguish pig iron from wrought iron.

2. What is a plain carbon steel and write a note on mild steels.

3. Explain the following:

(i)   Medium carbon steels

(ii)   High carbon steels

4. What is the purpose of adding alloying elements to steel?

5. Explain the influence of various alloying elements in steel.

6. What are special purpose steels?

7. Explain the following:

    (i)   High speed tool steels
    (ii)   High chromium steels

8. Briefly describe the types of stainless steels.

9. Write notes on the following:

    (i)   Elinvar
    (ii)   perm alloy
    (iii) Vlaue steels

10. Differentiate between gray cast iron and white cast iron.

11. Write note on different types of cast iron.

12. Explain in detail the production of spheroidal graphite cast iron.

13. What is malleable cast iron ? Explain with diagram the heat treatment cycle used for the production of malleable cast iron.

14. Write brief notes on

    (a) Inoculated cast iron
    (b) Ally cast irons

2   NON FERROUS ALLOYS

Metallic materials are divided into two large groups namely ferrous and non ferrous. The iron based materials are called as ferrous materials and the materials which have same element other than iron as the principal constituent are called as non ferrous materials. The most extensively used non ferrous materials are alloys of copper, aluminium, magnesium, nickel, zinc, tin, lead etc. The composition, properties and applications of some important non ferrous alloys are discussed below.

2.1 COPPER ALLOYS

Copper alloys may be classified into four groups as follows:

(a) Brasses

(b) Bronzes

(c) Cupronickels

(d) Nickel silvers

2.1.1 Brasses

Brasses are alloys of copper and zinc and containing more than 5 per cent zinc. Brasses are broadly classified as :

(a) $\alpha-$ brasses

(b) $(\alpha + \beta)$ brasses

2.1.1.1 α – *Brasses*

α - brasses contain zinc upto 36%. α - brasses having 5 to 20 per cent zinc are reddish in colour and are called as red brasses. α - brasses having zinc content between 20 and 36 per cent are yellow in colour and are called yellow brasses.

Commercial Brass : Commercial brass contains 90 per cent copper and 10 per cent zinc. It is stronger, harder than pure copper and is used for rivets, screws, jewellary.

Cartridge Brass : Cartridge brass contains 70 per cent copper and 30 per cent zinc. It has excellent deep drawing property and is used for making cartridge cases, house hold articles, radiator fins, lamp fixtures etc.

Admirality Brass : Admirality brass contains 70 per cent copper, 29 per cent zinc and 1 per cent tin. It has superior corrosion resistance than that of ordinary brass and is extensively used for propellers and marine works.

Aluminium Brass : Aluminium brass contains 76 per cent copper, 22 per cent zinc and 2 per cent aluminium. It has better corrosion resistance than admirality brass and hence extensively used for marine works.

2.1.1.2 $(\alpha + \beta)$ *Brasses*

$(\alpha + \beta)$ brasses contain copper between 54 and 62 per cent.

Muntz Metal : The most widely used $(\alpha + \beta)$ brass is muntz metal which contains 60 per cent copper and 40 per cent zinc. It has high strength and excellent hot working properties and is extensively used for marine fittings, condenser heads, radiator cores, springs, chains etc.

Naval Brass : Naval brass contains 60 per cent copper, 39 per cent zinc and 1 per cent tin. It has high corrosion and abrasion resistances and is widely used for condenser plates, propeller shafts and in marine works.

2.1.2 Bronzes

Alloys of copper with all other elements except zinc are called bronzes. The most important bronzes are alloys of copper and tin, aluminium, silicon or beryllium. These may also contain phosphorus, lead, zinc or nickel.

2.1.2.1 *Tin Bronzes or Phosphor Bronzes*

Tin bronzes are alloys of copper and tin and contain tin between 1 and 11 per cent. These are also called as phosphor bronzes since phosphorus invariably present as a deoxidizer in casting. The phosphorus content varies from 0.01 to 0.05 per cent.

These bronzes possess high strength, toughness, high corrosion resistance, low coefficient of friction and do not susceptible to season cracking. These are used for bushes, cotter pins, clutch disks, springs, taps, marine pumps etc. Tin bronzes have good castability and are widely used in the foundry.

Gun Metal : Gun metal contains 88 per cent copper, 10 per cent tin and 2 per cent zinc. It has considerable strength and toughness and resistance to sea water corrosion. It is used for bushes, nuts, hydraulic fittings, heavy load bearings, marine pumps etc.

2.1.2.2 *Aluminium Bronzes*

Commercial aluminium bronzes contain aluminium between 4 and 11 per cent. Aluminium bronzes containing up to 7.5 per cent aluminium are generally single phase alloys, while those containing between 7.5 and 11 per cent aluminium are two-phase alloys.

The single phase aluminium bronzes have good cold working properties, good strength and good corrosion resistance. These are used for corrosion resistant vessels, nuts, condenser tubes, bolts etc.

The properties of two phase aluminium bronzes can be improved by heat treatment. By controlling the composition and by heat treatment, the tensile strength of these alloys can be varied from 47 to 94 kg/sq.mm, with elongation gradually decreasing from 70 to 5 per cent. The heat treated aluminium bronzes are used for gears, propellers, pumps parts, bearings, bushings, drawing and forming dies.

2.1.2.3 *Silicon Bronzes*

Silicon bronzes contain 90 to 97 per cent copper, 1 to 4 per cent silicon and small amounts of zinc, iron and manganese. The thermal and electrical conductivity of these alloys is about 10 per cent of that of pure copper. Silicon bronzes have mechanical properties comparable to that of mild steel and corrosion resistance comparable to that of pure copper. These are used for storage vessels for chemical and gases, marine construction, bolts, nuts, rivets etc.

2.1.2.4 *Beryllium Bronzes*

Beryllium bronzes contain 1.5 to 2.25 per cent of beryllium. These alloys can be easily cast, can be easily hot or cold worked and can be easily welded. Like aluminium alloys, beryllium bronzes can be age hardened and the hardness obtained will vary between 200 to 400 Brinell, depending upon the aging time. And also tensile strengths upto 142 kg/sq.mm can be obtained by suitable heat treatment and cold working. The thermal and electrical conductivities of these alloys is comparable to that of pure copper.

Beryllium bronzes are used for diaphragms, springs, surgical and dental instruments, gears, watch parts, screws, bearings etc.

2.1.3 **Cupro Nickels**

The alloys of copper and nickel are called cupro nickels and contain upto 30 per cent nickel.

25% nickel alloys are used for utensils, coinage and in ships. These alloys can be easily hot or cold worked, soldered, brazed and welded.

30 per cent nickel alloy containing 1 per cent manganese and 1 per cent iron can be cast into small castings. These are used in pumps, valves, condenser tubes due to its good resistance to corrosion by sea water.

Cupro nickels are not susceptible to heat treatment. However, their properties can be altered by cold working. Cold working will increase the tensile strength from about 36 kg/sq.mm to a maximum of 66 kg/sq.mm and the elongation is gradually decreased from 50 per cent to about 5 per cent.

Constantan : Constantan is a copper alloy containing 40 per cent nickel, 1.5 per cent manganese and the rest copper. The main feature of constantan is its high electrical resistivity which remains almost constant upon variations in temperature. Constantan is used for rheostats, thermo couples and heating devices operating at moderately high temperatures.

2.1.4 Nickel Silvers (German Silver)

These are alloys of copper, nickel and zinc and contain 20 to 30 per cent nickel, 10 to 30 per cent zinc and the balance copper. The appearance of these alloys is similar to silver and possess good corrosion resistant characteristics. These are mainly used for utensils, costume jewellary, name plates, etc.

2.2 ALUMINIUM ALLOYS

Aluminium alloys are broadly classified into two groups:

    (a) Wrought aluminium alloys

    (b) Casting aluminium alloys

2.2.1 Wrought Aluminium Alloys

These alloys are further classified into:

    (a) Non heat treatable wrought aluminium alloys

    (b) Heat treatable wrought aluminium alloys

2.2.1.1 *Non Heat Treatable Wrought Aluminium Alloys*

Alloys which do not respond to heat treatment are called as non heat treatable alloys. They may be strain hardened. Examples of these alloys include aluminium-manganese and aluminium-magnesium alloys.

Aluminium-Manganese Alloys : The maximum solubility of manganese in the solid solution is 1.82 at the eutectic temperature of 1216°F. This alloy is strengthened by strain hardening and not by heat treatment. This alloy possess good formability, very good resistance to corrosion and good weldability. It is widely used for utensils, food and chemical handling equipment, gasoline and oil tanks, pressure vessels etc.

Aluminium-Magnesium Alloys: This alloy contains around 5 per cent magnesium with low silicon content. These alloys are not heat treatable and possess good resistance to corrosion, good weldability and moderate strength. These are widely used for aircraft fuel and oil lines, marine and welded structural applications etc.

2.2.1.2 *Heat Treatable Wrought Aluminium Alloys*

Alloys which respond to heat treatment and develop optimum properties are called as heat treatable alloys. Examples of these alloys include avial (Al- Mg- Si), aluminium-copper, and duralumins.

Avial (Al-Mg-Si) : The usual composition of this alloy is 0.45 - 0.9 per cent magnesium, 0.5 - 1.2 per cent silicon and 0.2 - 0.6 per cent copper. This alloy is heat treatable and the heat treatment consists in heating to 530 – 540 °C, quenching in water and subsequent natural ageing or artificial ageing at 150 – 160 °C for 12 to 16 hours. Immediately after quenching, ageing must be carried out, otherwise the strengthening effect is reduced. This alloy attains its full strength due to the precipitation of the hard, intermetallic compound Mg_2 Si after ageing.

Avial possess good weldability, corrosion resistance and immunity to stress corrosion cracking. Typical applications include furniture, aircraft landing mats and architectural applications.

Aluminium-Copper Alloys : Copper is one of the most important alloying elements of aluminium. The solid solubility of copper in aluminium decreases from 5.65 per cent at 548 °C to less than 0.25 per cent at room temperature. Normally in commercial alloys the copper content is limited to 4.5 per cent.

Copper forms the compound $Cu\ Al_2$ with aluminium. The mechanical properties of aluminium-copper alloys are improved after precipitation hardening due to the precipitation of theta phase ($Cu\ Al_2$) from the solid solution.

Aluminium-copper binary alloys have limited commercial applications and hence these alloys are usually combined with small percentages of other elements such as silicon, manganese, magnesium, zinc etc.

The corrosion resistance of aluminium is greatly decreased by the addition of copper and it is said to be the poorest of the conventional aluminium alloys. In order to improve corrosion resistance, aluminium - copper alloys are sandwiched between two pure aluminium sheets and rolled to produce a composite known as Alclad.

Typical applications include sheets, plates, forgings for structural and aircraft applicalions.

Duralumin Alloys : The typical chemical composition of duralumin is 4.3 - 4.9 per cent copper, 0.6 - 0.8 per cent magnesium, 0.6 - 0.8 per cent manganese. Manganese improves the corrosion resistance of duralumin.

The mechanical properties of duralumin are increased after heat treatment due to the precipitation of three phases such as $Al_2\,Cu\,Mg$, $Cu\,Al_2$ and $Al_6\,Cu\,Mg_4$.

Iron is a harmful impurity in duralumin alloys. By forming the compound $(Mn,\,Fe)\,Al_6$, iron considerably reduces the strength and ductility of duralumin. Hence iron content should never exceed 0.5 - 0.6 per cent. Duralumin is used for the manufacturing of various machine parts. Due to high ductility and good electrical conductivity, duralumin is used for making surgical instruments, cables, fluorescent tubes etc.

2.2.2 Cast Aluminium Alloys

The most important cast aluminium alloys are based on the aluminium - silicon system. These aluminium - silicon alloys are called as silumin alloys.

Silicon is the most important alloying element of cast aluminium alloys. It forms an eutectic with 11.7 per cent silicon and imparts high fluidity to the molten metal. Also, silicon reduces the solidification shrinkage (which is an important draw back of pure aluminium) to a considerable level from about 6 per cent for pure aluminium to about 3.8 per cent for 12 per cent silicon alloy.

Aluminium-silicon alloys offer excellent castability (high fluidity and low shrinkage), good corrosion resistance, high ductility and low specific gravity. These alloys can be strengthened by structural modification.

Aluminium - silicon alloys are widely used for intricate castings, aircraft industry, automotive industry, food handling equipment and marine fittings.

2.3 MAGNESIUM ALLOYS

The alloying elements added to magnesium are mainly aluminium, zinc, manganese and for special purposes tin, zirconium, cerium, thorium, silicon, calcium and beryllium. Copper, iron and nickel are the most undesirable impurities and must be kept at a minimum to provide the corrosion resistance in the alloy.

In most magnesium alloys, aluminium is the main alloying element ranging in amounts 3 to 10 per cent. The alloy becomes brittle if aluminium exceeds 10 per cent. Aluminium increases strength, hardening and castability of magnesium if added in the range of 3 to 10 per cent.

In magnesium base alloys, zinc is usually added together with aluminium in amounts upto about 3 per cent. If zinc exceeds 3 per cent, it produces porosity and brittleness due to the formation of the compound $Mg\ Zn_2$. Zinc effectively improves mechanical properties and also the casting properties of magnesium.

Manganese is added in magnesium - aluminium and magnesium - aluminium - zinc alloys in quantities less than 0.50 per cent. The main role of manganese is to improve corrosion resistance and weldability.

Addition of tin in magnesium - aluminium - manganese alloy, improves dllctility and forging properties. The maximum solubility of tin in magnesium is about 15% at 649 °C.

An addition of 0.6 to 0.85 per cent zirconium considerably refines the grain and increases the mechanical properties of magnesium alloys.

Addition of cerium to magnesium alloys increases the thermal conductivity and minimises the tendency towards hot shortness.

Addition of thorium greatly improves the elevated temperature properties of magnesium.

Addition of silicon makes magnesium more fluid and improves the casting properties.

Addition of calcium to magnesium greatly reduces oxidation losses during melting and heat treatment.

Addition of beryllium reduces the oxidizability of the molten alloy by forming an oxide film on the bath.

Magnesium alloys are generally used for making parts such as engines, gear boxes, flooring, seating etc., for aero planes, helicopters, missiles and satellites. Also these alloys are used for vehicle parts such as engines, transmission pumps and body panels.

2.4 NICKEL ALLOYS

The alloying elements added to nickel are mainly copper, iron, chromium, silicon, molybdenum, manganese and aluminium.

Monel : Monel is the most important of the nickel - copper alloys. It contains 67 per cent nickel, 30 per cent copper and small amounts of iron and manganese. Monel has high corrosion resistance to acids, alkalies, brines, water and food products. Monel is extensively used in the chemical, pharmaceutical, marine, electrical, textile fields. The various monel family alloys are R- Monel, K- Monel, H- Monel and S- Monel.

R- Monel possesses the same characteristics as Monel, but it contains high sulphur (0.035 per cent) to improve machanibility. This is mainly used for automatic screw machines.

K-Monel is an age-hardnable monel which contains about 3 per cent aluminium and possess corrosion resistance, similar to Monel, with high strength. This is mainly used for marine pump shafts, springs, aircraft instruments, ball bearings and safety tools.

H-Monel and S-Monel are casting alloys and contains 3 and 4 per cent silicon respectively. Both alloys have similar mechanical properties, but H-Monel has better machinability due to less silicon content (3 per cent). These alloys are mainly used for valves, seats, pump liners and impellers.

Hastelloy : These alloys contain about 60 per cent nickel, 20 to 30 per cent copper, 5 to 18 per cent iron and small quantities of manganese. These alloys possess excellent corrosion and oxidation resistance and extensively used for evaporators, reaction vessels, pipelines and fittings in the chemical industry.

Nichrome : This alloy contains 60 per cent nickel, 12 per cent chromium, 26 per cent iron and 2 per cent manganese. Nichrome shows good resistance to oxidation and heat and possess high melting point. It is widely used for making resistance coils, heating elements in stoves, electric irons and other house hold electrical appliances since it possess high electrical resistance.

Inconel : This alloy contains approximately 77 per cent nickel, 15 per cent chromium and 8 per cent iron. It is very ductile and shows excellent resistance towards corrosion and retains hardness even at 400 °C. This is widely used in carburizing boxes, retorts, muffles and thermocouple protection tubes.

Elinvar : This alloy contains 36 per cent nickel and 12 per cent chromium and possess zero thermoelastic coefficient i.e., the modulus of elasticity is invariable over a considerable range in temperature. It is used for hairsprings and balance wheels in watches and for similar parts in precision instruments.

Permalloys : These are nickel-iron alloys and contains nickel in the range of 78 per cent and possess very high magnetic permeability, low hysteresis losses and low electrical resistivity. These are used as loading coils in electrical communication circuits.

Alnico : The aluminium-nickel-cobalt alloys are called as alnico and contains 8 to 12 per cent aluminium, 14 to 28 per cent nickel, 5 to 35 per cent cobalt. These alloys have excellent magnetic properties and therefore extensively used as permanent magnets in motors, generators, radio speakers, microphones, telephone receivers and galvanometers.

2.5 LEAD ALLOYS

The alloying elements added to lead are mainly antimony and tin.

Antimony is generally added to lead to raise recrystallization temperature and to increase hardness and strength. Lead - antimony alloys contain 1 to 12 per cent antimony and are extensively used for storage battery plates, collapsible tubes and for building construction.

Solders : Solders are low melting point alloys of lead and tin. Soft solders contain 38 per cent lead, 31 per cent tin and 12% antimony. These are used for soldering electrical connections, for sealing tin cans and for joining lead pipes.

Silver soldiers contain 11 per cent silver, 16 per cent copper and 5% zinc and is used for brazing.

Type Metals : Lead-tin-antimony alloys are known as type metals and contains 75 per cent lead, 20 per cent antimony and 5 per cent tin. These are widely used in the printing industry.

Babbits : The alloys known as babbits or white metals are either lead based or tin based. A typical lead based white metal contains 80 per cent lead, 10 per cent antimony and 10 per cent tin and a typical tin based white metal contains 90 per cent tin, 5 per cent antimony and 5 per cent copper. Tin base white metals are more expensive but have better corrosion and wear resistance. These are widely used for making engine bearings.

2.6 MODEL QUESTIONS

1. Explain different types of Brasses.

2. Write short notes on the following:
    (i)   cartridge brass
    (ii) Admirality brass
    (iii) Naval brass

3. What are the applications and properties of phosphor bronze?

4. Write Short notes on the following:
    (i)   Gun metal
    (ii) Aluminium bronze
    (iii) Silicon bronze

5. Briefly explain the properties, composition and applications of various aluminium alloys.

6. Write short notes on the following.
   (i)   Monel
   (ii)   Nichrome
   (iii) Alnico
   (iv)   Babbits

3 PLASTIC DEFORMATION OF METALS

The changes that occur in metals or alloys under the action of applied forces is called as deformation of metals. By deforming metals or alloys, a desired shape can be achieved. The various deformation processes are rolling, forging, drawing, extrusion etc.

The imperfections in crystals which plays an important role in understanding the concept deformation of metals are discussed in detail before discussing the deformation of metals.

3.1 IMPERFECTIONS IN METAL CRYSTALS

A metallic crystal is fundamentally composed of atoms which are arranged in a specific geometric pattern which is repetitive in three dimensions. Crystals are of two types such as ideal crystals and real crystals.

An *ideal crystal* is the one in which atoms are arranged in a perfectly regular manner and is perfect without any defects.

Real crystals are not perfect and contain defects such as lattice distortion and various imperfections.

The presence of these imperfections or defects in metallic crystals impairs the physical and mechanical properties of metals and alloys.

3.1.1 Types of Imperfections or Defects

The various types of crystal imperfections are :

1. Point defects

    (a) Vacancies

    (b) Interstitial atoms

    (c) Impurities

    (d) Frankel defect

    (e) Schottky defect

2. Line defects : Dislocations

3. Surface or Grain boundaries defects

    (a) Grain boundaries

    (b) Tilt boundaries

    (c) Twin boundaries

4. Volume defects : Stacking faults

3.1.1.1 *Point Defects*

A point imperfection is a very localized interruption in the regular arrangement of a lattice, e.g., a vacant site. At equilibrium, the fraction of lattice sites that are vacant at a given temperature can be determined from the equation,

$$\frac{n}{N} = e^{-E_s/kT}$$

where

    n is the number of vacant sites

    N is the total number of atomic sites per cubic metre

    E_s is the energy required to move an atom from the interior of a crystal to its surface to create a vacancy in the lattice.

    K is the Boltzmann constant and

    T is the absolute temperature

Vacancies : Vacancies are empty atomic sites in cryslal lattice. Vacancies may arise in a crystal lattice due to imperfect packing during solidification or crystallization and they may also arise due to thermal vibrations of atoms at elevated temperatures since at higher temperatures the probability of atoms jumping out of their lattice position is more. When the density of vacancies becomes relatively large it is possible for them to cluster together to form voids in the lattice.

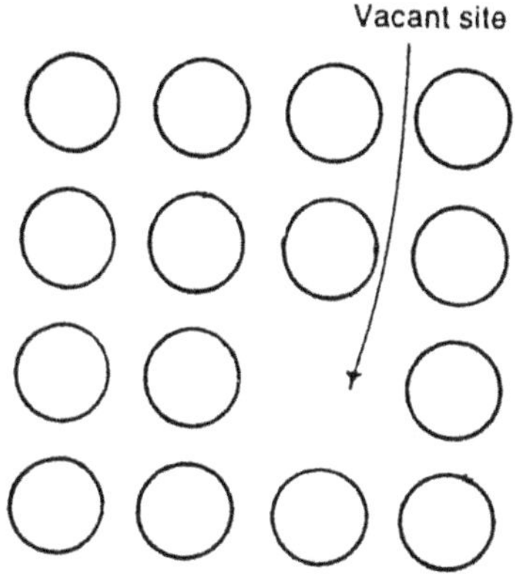

Fig. 3.1 *Vacancy Defect.*

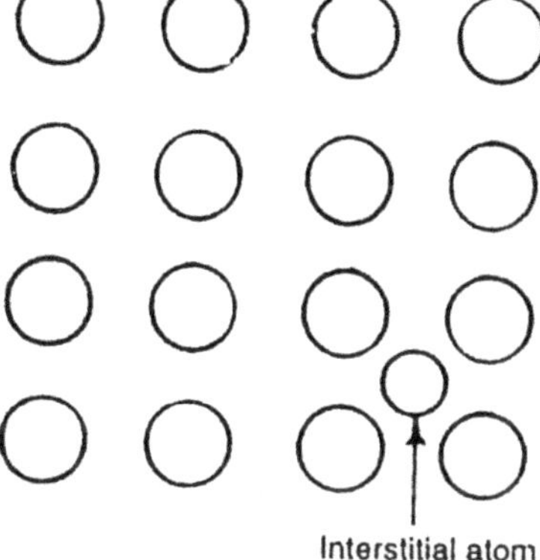

Fig. 3.2 *Interstitial Defect.*

Figure 3.1 shows the vacancy defect. The lattice planes will be distorted when the atoms surrounding a vacancy comes close together.

Interstitial (atom) defect : Interstitial defect arises when an atom occupies an interstitial position i.e., between the atoms in the lattice of the ideal crystal.

Figure 3.2 shows the interstitial defect. The interstitial atom may be either a normal atom of the crystal or a foreign atom. Interstitial defect occurs if the atomic packing factor of the crystal structure is low.

The interstitial defect produces lattice distortion since to occupy an interstitial position in a perfect crystal (ideal crystal), the atom tends to push the surrounding atoms farther apart. If the interstitial is smaller than the rest of the atoms in the crystal, the lattice distortions produced will be negligible.

Impurities : Impurities are small particles embedded in the structure such as slag inclusions in metals or foreign atoms in the lattice structure.

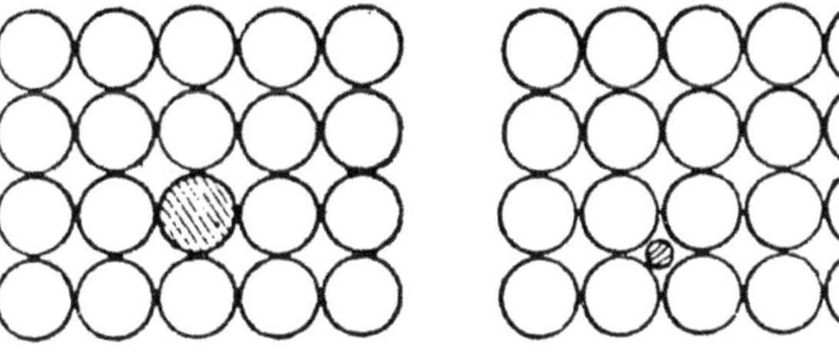

Fig. 3.3 *Impurities.*

Impurity atoms are of two types either *substitutional* or *interstitial atoms* (Fig. 3.3). Substitutional impurity atom refers to that atom which substitutes for or replaces a parent atom in the crystal. For example in brass, zinc is a substitutional atom in the copper lattice.

Interstitial impurity atom refers to that atom which occupy an interstitial position i.e., between the atoms of the ideal crystal. For example, the presence of carbon in iron where carbon occupies an interstitial position in the iron lattice structure without displacing any of the parent atoms of iron lattice from their positions.

Considerable distortion of lattice structure occurs due to the presence of impurities in the crystal or lattice structure.

Frankel Defect : Frankel defect is observed in ionic crystal. Frankel defect arises when an ion moves from a regular site to an interstitial site. Cations are generally the smaller ions and hence they move into the void space or interstial position. Where as anions do not enter into the void space because they can not be accommodated in the void space since the void space is too smaller in size than anions size. Silver halides and CaF_2 are the examples of Frankel, type.

Schottky Defect : Schottky defect is also observed in ionic crystals. Schottky defect arises due to vacancies in crystals. A pair of vacant ion sites i.e., one cation and one anion which are missed in an ionic crystal is called a schottky defect. Schottky defect is dominant in alkali halides.

3.1.1.2   *Line Defects - Dislocations*

The most important two dimensional or line defect is the dislocation. The explanation of the discrepancy between the computed and the real yield stresses lies in the fact that real crystals are not perfect crystals but contain defects. The type of defect which makes possible an explanation of the low yield point of crystals is the dislocation. Dislocations were originally postulated because they were able to explain the low yield points of metals and for many years there was no independent experimental evidence proving their existence. In the past few years, however experiments have been performed that proved the existence of dislocations beyond a reasonable doubt.

A dislocation may be defined as a disturbed region between two substantially perfect parts of a crystal.

There are two basic types of dislocations :

(a) Edge dislocation

(b) Screw dislocation

Edge Dislocation

Fig. 3.4 shows an edge dislocation in a simple cubic crystal. The line marked as SP represents the slip plane of the crystal and the dashed lines represents the crystal planes.

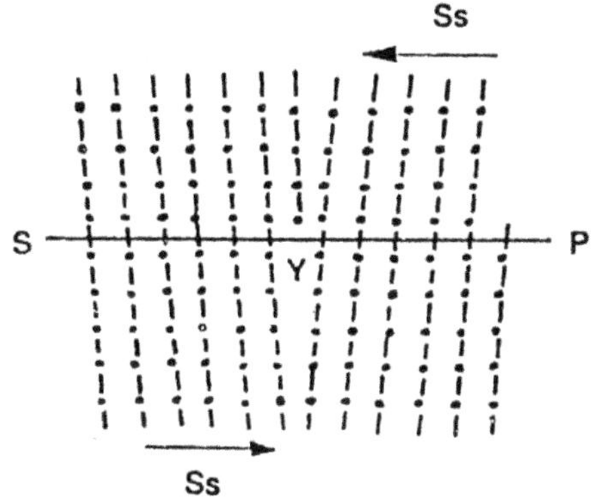

Fig. 3.4 An edge dislocation.

The plane marked 'X' at the top of the figure ends at point 'Y' on the slip plane, where as the pl nes on either side of 'X' run continuously from the top to the bottom of the figure. In such a case where a lattice plane ends inside a crystal, an edge dislocation is the result.

The edge dislocation shown in figure has an incomplete plane which lies above the slip plane. Such an edge dislocation is called as positive edge dislocation and is represented by the symbol '⊥' where the vertical line represents the incomplete plane and the horizontal line represents the slip plane. It is also possible to have the incomplete plane below the slip plane which can be represented by the symbol ' ⊤.'

Screw Dislocation

A screw dislocation is shown schematically in Fig. 3.5. The upper front portion of the crystal has been sheared by one atomic distance to the left relative to the lower front portion. No slip has taken place to the rear portion of the line DC and therefore DC is a dislocation line. The plane ABCD is the slip plane. The designation screw for this lattice defect is derived from the fact that the lattice planes of the crystal spiral the dislocation line DC.

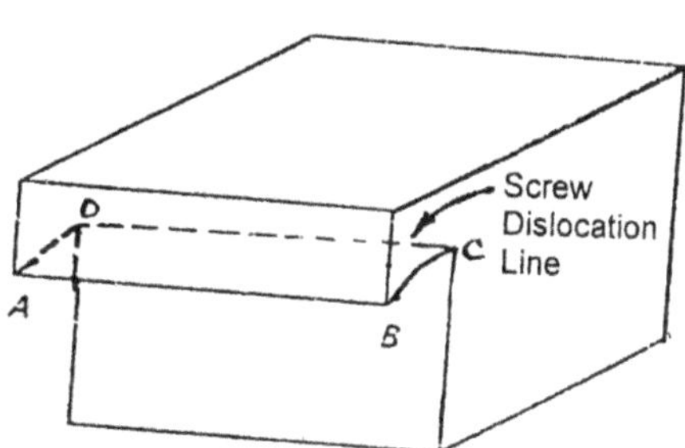

Fig. 3.5 Representation of a screw dislocation.

Screw dislocations are also of two types (i) *left hand screw* and (ii) *right hand screw* dislocation. In left hand screw dislocations the lattice planes spiral the dislocation line DC like a left hand screw. In the right hand screw dislocation, the lattice planes spiral the dislocation line DC like a right hand screw.

The Burgers Vector : The Burgers vector 'b' is the vector which defines the magnitude and dlrection of slip. Therefore it is the most characteristic feature of a dislocation. If the Burgers vector and the orientation of the dislocation line are known then the dislocation is completely described.

Fig. 3.6 shows a method of determining the Burgers Vector applied to a positive edge dislocation. It is first necessary to choose arbitrarily a positive direction for the dislocation. In the present case it can be assumed a counter clockwise direction. Fig. 3.6(A) shows a counter clockwise circuit, of atom to atom steps in a perfect crystal, closes i.e., the end point meets the starting point. On the other hand, when the same step by step circuit is made (Fig. 3.6B) around the dislocation in an imperfect crystal, the end point of the circuit does not meet the starting point. The vector 'b' connecting the end point with the starting point is the Burger's Vector of the dislocation. The procedure can a be used to find the Burgers vector of any dislocation.

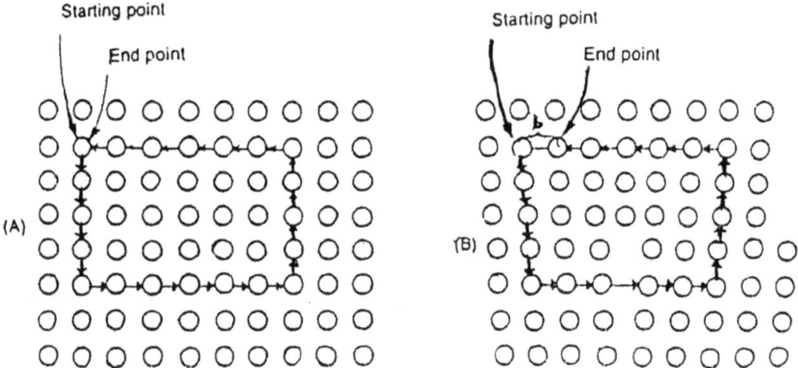

Fig. 3.6 *The Burgers circuit for an edge dislocation. (A) Perfect crystal and (B) Crystal with dislocation.*

Fig. 3.7 shows the Burgers circuit for screw dislocation. Fig. 3.7 (A) shows the circuit for perfect crystal and Fig. (B) shows the same circuit for an imperfect crystal containing a screw dislocation. Based on the concept of Burgers Vector, certain characteristics of edge and screw dislocations are as follows:

(a) An edge dislocation lies perpendicular to its Burgers Vector where as a screw dislocation lies parallel to its Burgers Vector

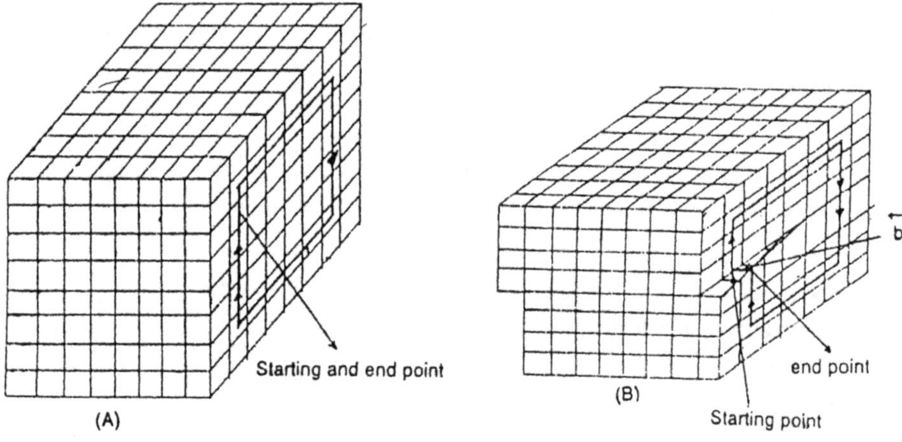

Fig. 3.7 *The Burgers circuit for a screw dislocation.(A) Perfect crystal (B) Crystal with dislocation.*

(b) An edge dislocation moves (in its slip plane) in the direction of the Burgers Vector (slip direction) where as a screw dislocation moves (in the slip plane) in a direction perpendicular to the Burgers Vector (slip direction)

The Frank-Read Source - Dislocation Generator : The generation of dislocations can be explained by the Frank-Read mechanism which involves the Frank-Read source and its operation

A Frank-Read source consists of a dislocation line fixed at nodes. Consider a positive edge dislocation segment x y (Fig. 3.8). Under applied shear stress the dislocation segment (line) bows out and move to the left, causing the line to form an arc with its ends fixed at end point x and y. This arc is shown in figure 3.8 by the symbol 'a'. Further increase in applied stress causes the curved dislocation to expand to the successive positions 'b' and 'c'. At 'c' the loop intersects itself at point 'm' and annihilate each other at the point of contact 'm' and breaks the dislocation into two segments marked 'd', one of which is circular and expands to the surface of the crystal, producing a shear of one atomic distance. The other component

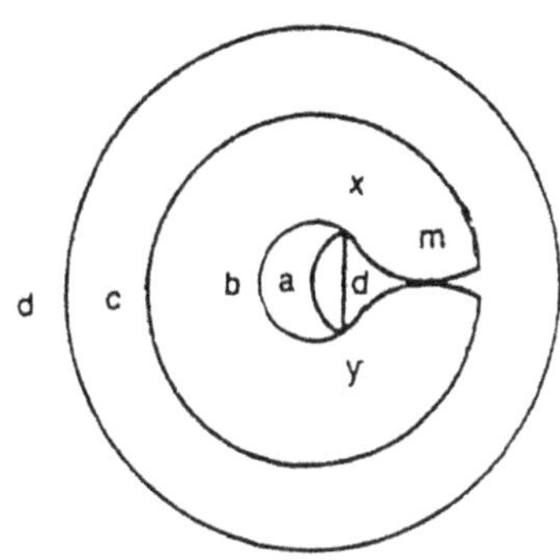

Fig. 3.8 Frank-Read source.

remains as a regenerated positive edge dislocation line lying between points x and y, where it is in a position to repeat the cycle. Many dislocation loops can be generated in this way on the same slip plane. This type of dislocation generator is called as Frank-Read source.

Nucleation of Dislocation : The existence of Frank-Read source is indicated by the experiments. But from the recent experimental evidence it can be stated that dislocations can also be formed without the aid of Frank-Read or similar sources. This clearly shows that if dislocations are not formed by dislocation generators then a nucleation process must have created dislocations. Creation of dislocations by nucleation process can be in two ways, *homogeneously* or *heterogeneously*. Homogeneous nucleation of dislocations means that dislocations are formed by the action of stress alone. Heterogenous nucleation of dislocations means that dislocations are formed with the help of crystal defects such as the presence of impurity particles in the crystal. The defects in the crystal lowers the stress required to form dislocations and causes the formation of dislocations much easier. On the other hand homogeneous nucleation of dislocations require very high stresses, theoretically of the order of 1/30 of the shear modulus of a crystal. Theoretically the shear modulus of a crystal is about 10^6 to 10^7 Psi and hence the stress required to form dislocations should be 1/30th of theoretical stress i.e., about 10^5 Psi. But practically at a shear stress of around 100 Psi, the metal crystals start deforming. This clearly indicates that dislocations must be nucleated heterogeneously if dislocations are not formed by Frank-Read source.

Climb of Edge Dislocation : An edge dislocation can move by a method which is fundamentally different from slip. This method is called as dislocation climb and involves motion of the dislocation in a direction perpendicular to the slip plane. Dislocation climb occurs by the diffusion of vacancies or interstitials, to or away from the site of dislocation. Since climb is diffusion controlled it is thermally activated and occurs more readily at elevated tcmperature.

Fig. 3.9 shows a positive climb of an edge dislocation. Figure shows a vacancy or vacant site just to the right of atom 'a'. If atom 'a' jumps into the vacancy the edge of the dislocation loses one atom as shown in Figure 3.9 and the edge dislocation will climb one atomic distance in a direction perpendicular to the slip plane. This type of climb is called as a positve climb and results in a decrease in size of the extra half plane.

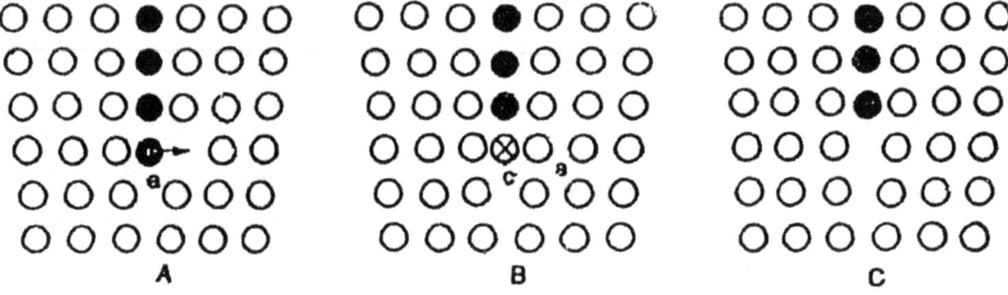

Fig. 3.9 Positive climb of an edge dislocation.

Negative climb is opposite to that of positive climb. In negative climb instead of shrinking, the extra plane grows in size. Fig. 3.10 shows the mechanism of negative climb of an edge dislocation.

To produce a negative climb atoms must be added to the extra plane of atoms. This can occur when the atoms from the surrounding lattice plane join the extra half plane and this creates a flux of vacancies which moves off into the crystal. As shown in Figures 3.10 (A) and (B), atom 'a' moves to the left and joins the extra plane leaving a vacancy to its right there by the extra plane grows in size producing a negative dislocation climb.

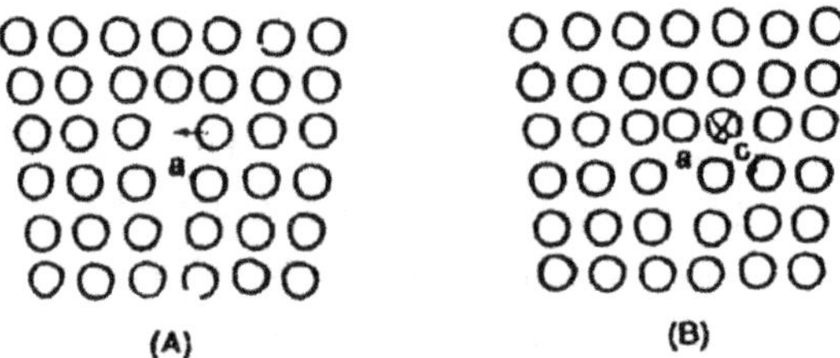

Fig. 3.10 Negative climb of an edge dislocation.

Positive climb causes the crystal to shrink in a direction perpendicular to the slip plane and therefore associated with a compressive strain and requires a compressive stress for promoting positive climb. On the other hand a tensile stress is required to promote negative climb. Hence there is a fundamental difference between the nature of the stress that produces slip (shear stress) and climb (compressive or tensile).

Cross Slip : Cross slip occurs in crystals when there are two or more slip planes with a common slip direction. During cross slip the screw dislocations shift from one slip plane to the other in order to avoid obstacles for their motion. Cross slip most readily occurs in those crystals which have a large number of slip systems such as in F.C.C. and H.C.C. crystal structures.

Force on Dislocation : When an external force of sufficient magnitude is applied to a crystal, movement of dislocations occur and produce slip. Hence a force should act on a dislocaton line to drive it forward.

Forces Between Dislocations : Dislocations of opposite sign on the same slip plane will attract each other, run together and destroy each other. For example, a positive and negative edge dislocation on the same slip plane would join together and form complete plane thereby eliminating the extra half plane of atoms. Hence the dislocation will disappear. Similarly, dislocations of like sign on the same slip plane will repel each other.

Intersection of Dislocations – Jogs : Fig. 3.11 (A&B) shows the intersection of two edge dislocations with Burgers Vectors at right angles to each other. An edge dislocation XY is moving on plane P_{xy} with Burgers vector 'b_1' and intersect the edge dislocation AD which is lying on plane P_{AD} with Burgers Vector 'b_2'. The intersection of these two edge dislocations produces a jog pp^1 (Fig. B) in the dislocation AD. The produced jog is parallel to 'b_1'. The resulting jog from this intersection of two edge dislocations has an edge orientation and hence it can glide with the rest of the dislocation. Jog usually forms when the Burgers vector of the intersecting dislocation is normal to the other dislocation line as in this case 'b_1' is normal to AD and jogs AD, where as 'b_2' is parallel to XY and no jog is formed.

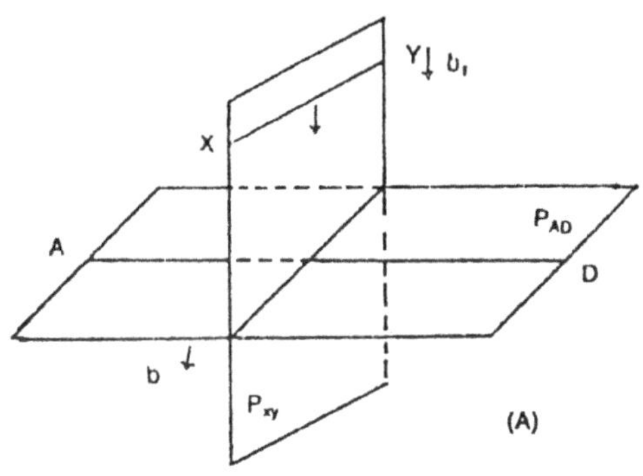

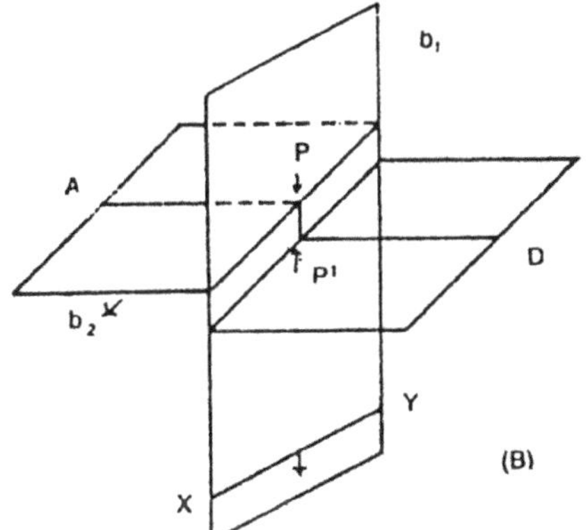

Fig. 3.11 *Intersection of two edge dislocations.*

The intersection of a serew dislocation and an edge dislocation produces a jog with an edge orientation on the edge dislocation and a kink (kinks are jogs which lie on the slip plane instead of normal to it and are unstable since during glide they can line up and annihilate) with an edge orientation on the screw dislocation. The intersection of two screw dislocations produces jogs of edge orientation in both screw dislocations. This is the most important type of intersection from the view point of plastic deformation.

The jogs produced by the intersection of two edge dislocations are able to glide readily because they lie in the slip planes of the original dislocations where as the motion of serew dislocations will be impeded by jogs.

Dislocation pile ups

Pile up of dislocations occur on slip planes at obstaclcs such as grain boundaries, second phase particles or sessile dislocations (immobile dislocations). The leading dislocation in the pile up is acted on not only by the applied shear stress but also by the inter-action force with the other dislocations in the pile up. Hence a high concentration of stress develops on the leading dislocation in the pile up. The stress on the leading dislocation in the pile up approaches the theoretical shear stress of the crystal, when many dislocations are piled up. This high stress can either initiate yielding on the other side of the obstacle or it can nucleate a crack at the obstacle.

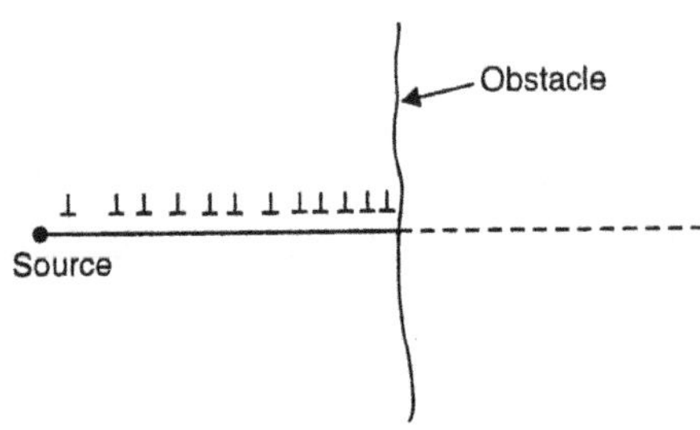

Fig. 3.12 *Dislocation pile up at an obstable.*

The number of dislocations which can be supported by an obstacle will depend on the factors such as, the type of obstacle, the orientation relationship between the slip plane and the structural features at the obstacle, the material and the temperature. Break down of the barrier can occur by slip on a new plane, by climb of dislocations around the obstacle or by the generation of high tensile stresses which produces a crack.

3.1.1.3 *Surface Defects*

Surface defects may include grain boundary, tilt boundary, twin boundary etc.

Grain Boundaries : Grain boundary is formed when two growing grain surfaces meet each other and these grain boundaries separate crystals or grains of different orientation in poly crystalline materials. The orientation difference between neighbouring grains is usually greater than 10 - 15° and for this reason these grain boundaries are called as high angle grain boundaries. In grain boundaries the atomic packing is imperfect and most of the atoms at the boundaries are in highly strained and distorted positions.

***Tilt Boundaries* :** Tilt boundaries are formed by edge dislocations and are regarded as an array of edge dislocations located one above the other. When the orientation difference between two crystals is less than 10°, the distortion in the boundary is not very drastic and these boundaries are called low angle tilt boundaries.

***Twist Boundaries* :** Another low angle boundaries formed by screw dislocations are called as twist boundaries.

***Twin Boundaries* :** Another type of surface imperfections are twin boundaries. A twin boundary separates two parts of crystal having the same orientation.

3.1.1.4 *Volume Defects*

Volume defects are stacking faults which are created by a fault in the stacking sequence of close packed atomic planes in crystals such as FCC and HCP.

Consider for example the stacking arrangement in an FCC crystal ABC ABC ABC ABC.

If an A plane indicated by an arrow is missing then the stacking sequence becomes ABC BC ABC ABC.

Which is a surface defect and is called as stacking fault. Stacking faults are more frequently found in deformed metals.

3.2 DEFORMATION OF METALS

Metals or alloys get deformed when they are stressed and the deformation causes the change in dimensions.

The deformation of metals is of two types :

(a) *Elastic deformation* : Elastic deformation is a temporary deformation which disappears after the deforming load is removed.

(b) *Plastic deformation* : Plastic deformation is a permanent deformation which remains even after the deforming load is removed

3.2.1 Elastic Deformation

When a piece of metal is subjected to a stress of a low magnitude a temporary deformation of the crystals takes place by displacing the atoms from their original positions. But when the deforming stress is removed, the atoms return to their original positions and the crystal recovers its original shape.

Fig. 3.13 shows the influence of tensile load on the atoms (the circles in figures) of the unit cell. Fig. 3.13(a) shows the front face of the F.C.C. unit cell before loading and Fig. 3.13(b) shows the same unit cell which is subjected to a tensile load and it can be observed that there is a slight elongation of the unit cell in the directions of tensile load. Fig. 3.13(c) shows the same unit cell which returned back to the original shape after the removal of tensile load.

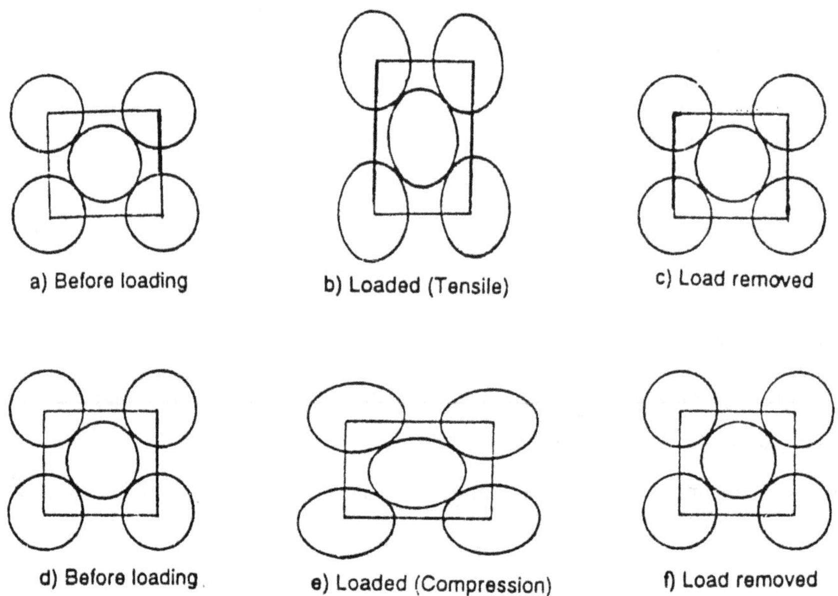

Fig. 3.13 *Elastic Deformation.*

Fig. 3.13 (d) shows the front face of the F.C.C. unit cell before loading and Fig. 3.13 (e) shows the same unit cell which is subjected to a compressive load and it can be observed that there is a slight contraction of the unit cell in the direction of the compressive load. Fig. 3.13 (f) shows the same unit cell which returned back to the original shape after the removal of compressive load.

Similarly the metals which are subjected to shear stresses produces shear strain i.e., displacement of one plane of atoms relative to the adjacent plane of atoms.

It is observed that when the elastic deformation occurs, the strain is nearly proportional to the applied stress and the ratio between stress and strain is known as Young's modulus of elasticity E, where

$$E = \frac{\text{stress}}{\text{strain}}$$

Young's modulus of elasticity is an important characteristic of a metal since it gives an idea about the elasticity of a metal.

3.2.2 Plastic Deformation

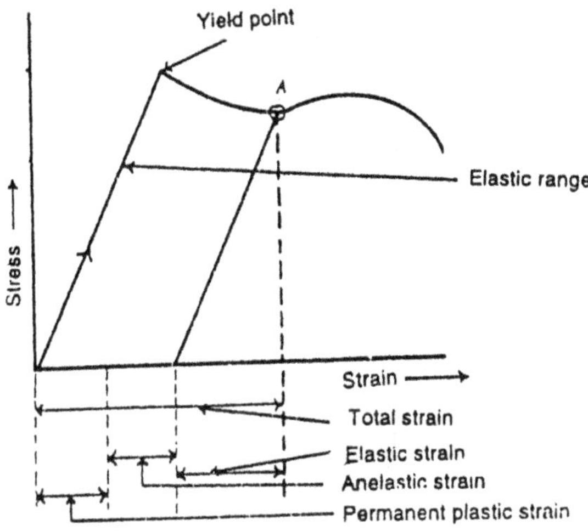

Fig. 3.14 *Permanent plastic strain.*

When the stresses in the metal specimen crosses the elastic limit, the specimen gets deformed permanently. This permanent deformation is called as the plastic deformation and causes the distortion of the crystal structure which is irreversible.

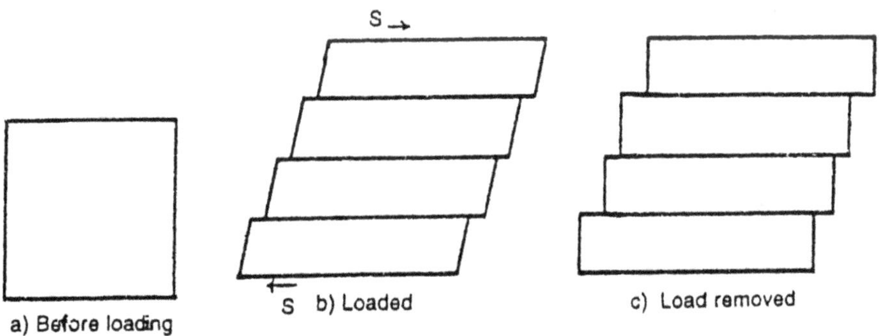

Fig. 3.15 *Plastic deformation.*

If the deforming load after it crosses the elastic limit is removed, the specimen does not come back to its original shape and there always remain a permanent plastic strain in the specimen which is shown in Fig. 3.14 and 3.15. As shown in the figure when the deforming load is suddenly removed from a plastically deformed specimen at point A, the elastic strain $\left(\dfrac{stress}{E}\right)$ is removed from the specimen immediately and the an elastic strain disappear slowly after the load has been removed but the plastic strain (deformation) remains in the specimen permanently.

Plastic deformation plays a vital role in metal shaping processes such as drawing, forging, bending, extrusion, stamping, rolling etc.

3.2.3 Mechanisms of Plastic Deformation

There are two important mechanisms of plastic deformation for single crystals. They are

(a)   Plastic deformation by slip, and

(b)   Plastic deformation by twinning.

These two mechanisms occur by pure shear stresses.

3.2.3.1 *Plastic Deformation by Slip*

The mechanism of plastic deformation in metals by sliding or moving or gliding of blocks of the crystal over one another along certain definite crystallographic planes called slip planes is known as slip.

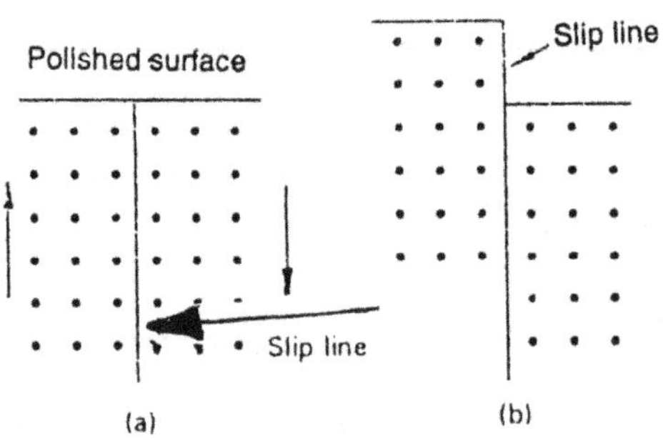

Fig. 3.16 *An idea about slip.*

Fig. 3.16 shows the mechanism of slip. Fig. 3.16 shows a metal cube with a top polished surface which is subjected to a shear stress. Slip occurs when the shear stress exceeds a critical value and the atoms move an integral number of atomic distances along the slip plane. Slip results in visible step on the polished surface of the crystal and this step appears as a line which is called a slip line when viewed the polished surface from above with a electron microscope.

Slip occurs most readily in specific directions on certain crystallographic planes. It is generally found that slip planes are the planes of greatest atomic density i.e., having highest number of atoms per unit area and slip direction is the closest packed direction. The slip planes are also the most widely spaced planes in the crystal structure, the resistance to slip will be generally less for these planes than for any other set of planes. The slip plane together with the slip direction that lies on it constitutes the slip system. The slip system for different crystal structures are given below:

Crystal structure	Slip plane	Slip direction
F.C.C.	(111)	< 110 >
B.C.C.	(110)	< 111 >
H.C.P.	(0001)	< 1120 >

If there are more than one set of slip planes, the slip starts on the set along which there is the maximum shear stress. There are 12 sip systems in F.C.C. crystals where as H.C.P. crystals have only 3 slip systems.

3.2.3.2 Critical Resolved Shear Stress for Slip

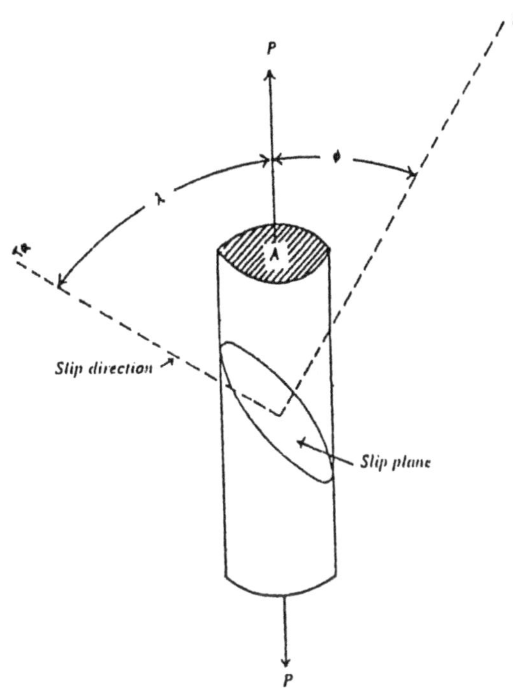

Fig. 3.17 Diagram for calculating critical resolved shear stress.

Slip begins when the shear stress on the slip plane in the slip direction reaches a threshold value called the critical resolved shear stress.

Critical resolved shear stress is a property of material which depends upon composition and temperature and does not depend on the crystal structure.

Consider a cylindrical single crystal with cross sectional area A (as shown in Fig. 3.17). The angle between the normal to the slip plane and the tensile axis is ϕ, and the angle which the slip direction makes with the tensile axis is λ. The area of the slip plane inclined at the angle ϕ will be A/Cosϕ.

The component of the axial load acting in the slip plane in the slip direction is P cos λ.

Therefore the critical resolved shear stress T_R is given by

$$T_R = \frac{\text{Load}}{\text{Area}} = \frac{P \cos \lambda}{A/\text{Cos } \phi} = \frac{P}{A} \text{ Cos } \phi \text{ Cos } \lambda \qquad \dots\dots 3.1$$

Eq. 3.1 is also known as Schmid's law.

Eq. 3.1 gives the shear stress resolved on the slip plane in the slip direction. This shear stress is a maximum when

$$\phi = \lambda = 45°, \text{ so that}$$

$$\tau_R = \frac{P}{A} \; \frac{1}{\sqrt{2}} \; \frac{1}{\sqrt{2}}$$

$$= 1/2. \; \frac{P}{A}$$

If the tension axis is normal to the slip plane ($\lambda = 90°$) or if it is parallel to the slip plane ($\phi = 90°$) the resolved shear stress is zero since cos $90° = 0$. Slip will not occur for these extreme orientations since there is no shear stress on the slip plane and crystals close to these orientations tends to fracture rather than slip.

The values of critical resolved shear stress of the principal slip systems at 20°C for different metals are given below :

Metal	Structure	Slip plane		Critical resolved stress (kg/cm^2)
		Slip system	Slip direction	
Iron	B.C.C.	(110)	< 111 >	280
Molybdenum	B.C.C.	(110)	< 111 >	730
Copper	F.C.C.	(111)	< 110 >	5
Aluminium	F.C.C.	(111)	< 110 >	8
Cobalt	H.C.P.	(0001)	<1120 >	67.5
Megnesium	H.C.P.	(0001)	< 1120 >	4.5

3.2.3.3 *Mechanism of Slip (dislocation theory)*

The modern concept of the mechanism of slip is that the slip occurs step by step by the movement of dislocations in the crystal. Slip usually originates at a point where dislocation exists in the crystal and the slip proceeds by the movement of the dislocation due to the applied shear stresses. A dislocation results ultimately in a surface step. The force required to produce slip is of the order of 1000 times less than the theoretically worked out for perfect crystal and this is due to the step by step movement of dislocations in imperfect crystals or real crystals.

3.2.3.4 *Plastic Deformation by Twinning*

The second important mechanism which causes the plastic deformation of metals is twinning. Twinning is a predominant mechanism particularly in H.C.P. metals. Twinning results when a portion of the crystal takes up an orientation that is related to the orientation of the rest of the untwinned lattice in a definite symmetrical way. The plane of symmetry between the two portions is called the twinning plane.

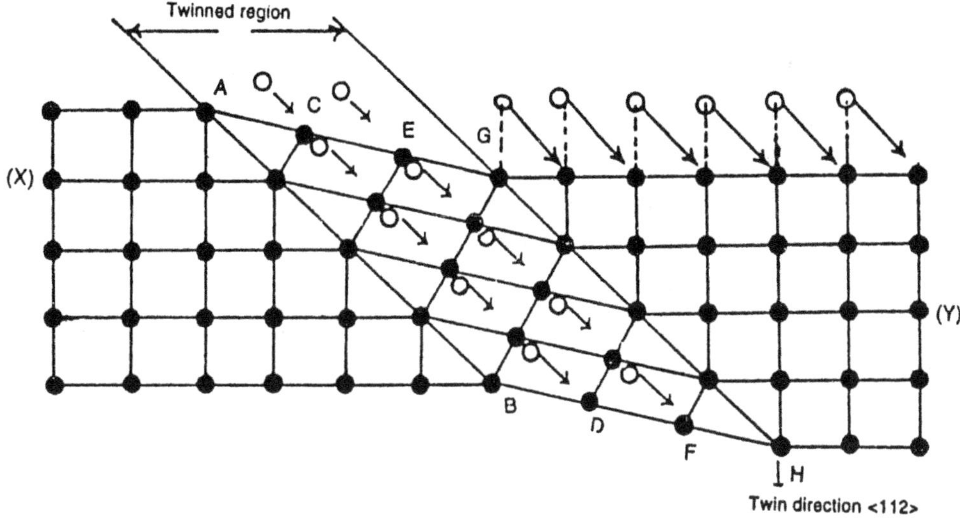

Fig. 3.18 *Twinning in a lattice.*

When stress is applied a crystal, whole blocks of atoms moves various distances along the slip plane during slip but when twinning occurs, each plane of atoms concerned moves definite distance in the same direction such that the extent of the movement of each plane is proportional to its distance from the twinning plane. The distance moved by atoms on each subsequent plane is more than that on the previous plane atoms by a given fraction of the unit atomic distance. As shown in the Fig. 3.18, the plane CD moves 1/3rd of an interatomic spacing whereas the plane EF moves 2/3rd of an interatomic spacing and the plane GH moves 3/3rds i.e., one interatomic spacing. The movement of these planes causes the formation of a twinned region. As shown in the figure, the crystal is separated into two parts by the twinned region. These two parts are oriented in such a way that one forms the mirror image of the other with respect to the twin plane between them.

Plastic defonnation occurs by twinning due to an extensive change in shape of the crystal and also due to bringing the planes of potential slip into more favourable orientation with respect to the applied stress axis so that additional slip can take place.

Twinning occurs in a definite direction on a specific crystallographic plane for each crystal structure. The table below lists the common twin planes and twin directions for different crystal structures.

Table : Twin planes and twin directions.

Crystal Structure	Typical Examples	Twin plane	Twin direction
B.C.C.	Fe, Cr, V, Mo	(112)	111
F.C.C.	Ag, Au, Cu	(111)	112
H.C.P.	Zn, Cd, Mg, Ti	(1012)	1011

Whether or not there is a critical resolved shear stress for twinning is still not known. But if a crystal possess many possible slip systems. twinning is not a dominant deformation mechanism. When the slip systems are restricted or when something increases the critical resolved shear stress so that the twining stress is lower than the stress for slip then twinning occurs in a crystal. Twinning may occur due to impact, heat treatment or plastic deformation.

Types of Twins : There are two types of twins such as

(a) *Mechanical twins* **:** Mechanical twins are due to mechanical deformation and these are produced mostly in B.C.C. and H.C.P. metals under conditions of shock loading and decreased temperature. Mechanical twins formed in iron are called Neumann bands.

(b) *Annealing twins* **:** Annealing twins are the result of annealing following plastic deformation. Annealing twins usually occur in F.C.C. metals.

Difference Between Slip and Twinning

    (i)   Twins can form in a few micro seconds whereas slip takes several milliseconds.

    (ii) During slip all atoms in one block move the same distance whereas in twinning atoms in each successive plane within a block move different distances.

  (iii) During the formation of twins a loud sound called twin cry can be heard and this can not be heard for slip.

  (iv) Slip generally requires a lower shear stress than twinning.

    (v) Slip appears as thin lines when viewed under microscope where as twinning appears as broad lines.

  (vi) Slip requires a threshold value of stress called the critical resolved shear stress where as twinning requires no such threshold value of stress.

3.2.4 Deformation of Polycrystalline Materials

Commercial metal products are usually polycrystalline aggregates. These polycrystalline aggregates are made up of a tremendous number of small single crystals or grains. The individual small single crystals in polycrystalline aggregates can not deform like single crystals because in polycrystalline materials, the crystals are not free to deform since they are surrounded by other crystals and also there will be the influence of grain boundaries. Hence a larger stress is required for deformation of polycrystalline materials because each grain restricts the deformation of neighbouring grains.

The boundaries between grains plays a vital role during the plastic deformation of a polycrystalline aggregates. Grain boundaries acts as an obstacle to the motion of dislocations and pile up of dislocations occurs along the slip planes at the grain boundary. Under the influence of applied stresses more and more number of dislocations pile up and causes high shear stresses to develop at the leading dislocation in the pile up. These high stresses eventually become high enough to produce movement of dislocations in the neighbouring grain across the boundary.

Size of the grain has a tremendous influence in the early stages of deformation of polycrystalline materials. During the early stages of deformation grain boundary obstacles for dislocations are most effective and for deformation to proceed further these obstacles have to be broken. Thus yield strength of a polycrystalline material is more dependent on grain size than tensile strength. The yield stress increases with decrease in grain size because smaller the grains more is the number of grain boundaries and hence more obstacles for the movement of dislocations. Hall and Petch have related the yield stress to the grain size for a polycrystalline material by the following equation :

$$\sigma_0 = \sigma_i + kd^{-1/2}$$

where σ_0 is the yield stress

σ_i friction stress opposing the motion of a dislocation

k is a constant and depends on the extent to which dislocations are piled up at barriers

d is the average diameter of grain

A very common method of calculating grain size from the ASTM specification for grain size is as follows:

The ASTM number 'n' is related to N by the relationship

$$N = 2^{n-1}$$

where N is the number of grains per square inch (645 mm^2) at a magnification of 100 X.

For example grain diameter for ASTM 2 corresponds $2^{2-1} = 2$ grains per square inch at a magnification of 100 X (or) $2 \times \dfrac{10^4}{645} = 30.1$ grains per mm^2 without any magnification.

i.e., $1 \text{ mm}^2 = 30.1 \text{ grains}$

$1 \text{ mm} = \sqrt{30.1} \text{ grains}$

or $1 \text{ grain size (diameter)} = \dfrac{1}{\sqrt{30.1}} = 0.18 \text{ mm}$

The grain size or grain diameter for the specification of ASTM 2 is equal to 0.18 mm.

3.2.4.1 *Deformation Process of Polycrystalline Material*

When a polycrystalline material is subjected to an applied stress the material behaves perfectly elastic in the first stage since all the crystals are deformed only elastically. In the second stage only a few crystals undergoes plastic deformation by crossing the elastic limit. At this stage if the load is removed, the plastically deformed crystals slowly comes back to their original shape due to the stresses that are arised between the elastically and plastically deformed crystals. This gradual return is called the elastic after effect which is observed only in polycrystalline materials and not in single crystals. In the third stage majority of the crystals crosses the elastic limit and deform plastically which causes permanent deformation in the material. And in the last stage, all the crystals deform plastically and results in yielding of the material.

3.3 STRAIN HARDENING OR WORK HARDENING

Work hardening or strain hardening is a phenomenon which results in an increase in hardness and strength of a metal subjected to plastic deformation. The shear stress required to produce slip increase continuously with shear strain. The increase in the stress required to cause slip because of previous plastic deformation is called as work hardening or strain hardening.

As a means of improving the useful mechanical properties such as strength and hardness, strain hardening is commonly employed for pure metals and alloys. Strain hardening however reduces ductility and plasticity.

3.3.1 Principle of Work Hardening

Fig. 3.19 shows the principle of work hardening. When a metal specimen is subjected to a stress, the strain increases with stress and the curve reaches the point A in the plastic range. If at point A, the specimen is unloaded, the strain does not recover along the original path AO but moves along the path AB. Then if the specimen is reloaded immediately without lapse of time, the curve again rises from B to A, but follows another path and reaches the point C, and if the loading is continued it will follow the curvature ACD. At point A, if the specimen is not unloaded, the stress-strain curve would have followed the dotted path AD^1. Comparing the paths ACD and AD^1 it can be concluded that the cold working (plastic deformation) has increased the yield strength and ultimate strength of the metal since S_2 is greater than S_1.

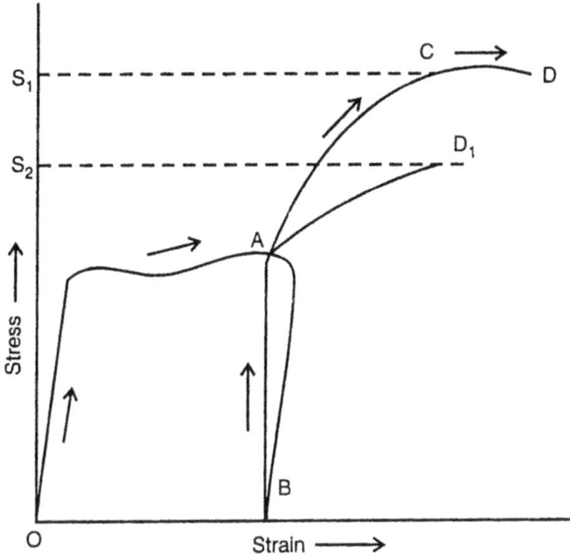

Fig. 3.19 *Work hardening.*

3.3.2 Theory of Work Hardening

Several theories are put forward to explain the phenomenon of work hardening. All these theories says that work hardening is due to the increased resistance to the motion of dislocations inside the crystal when the metal has been subjected to cold working (plastic deformation).

Taylor's classical theory of work hardening is based on dislocations which have been arrested inside a crystal. He assumes that some dislocations become struck inside the crystal and act as sources of internal stress which oppose the movement of other dislocations i.e., work hardening arises due to the interactions between dislocations. Hence the stress (τ) required, to move a dislocation in the stress field of other dislocations surrounding it, will have to be increased for further plastic deformation to occur.

Taylor has given the following relation between stress (τ) and strain ε ,

$$\tau = K.G. \ (\varepsilon \frac{b}{L})^{1/2}$$

where τ is the stress

               K is a constant

G is the shear modulus

ε is the strain

b is the Burgers vector

L is the distance to which a positive and a negative dislocation separate before being stopped.

3.3.3 Stages of Work Hardening

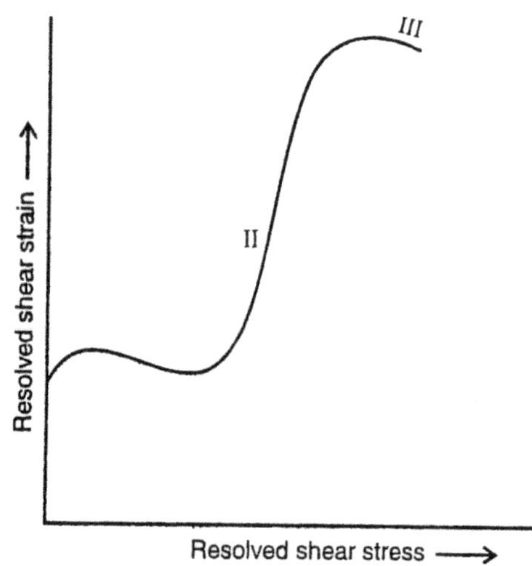

Fig. 3.20 Stress-strain curve showing three stages of work hardening.

Fig. 3.20 shows the stress-strain curve of an F.C.C. crystal showing three stages of work hardening which are observed experimentally by Seeger.

Stage I : In the first stage there is very little strain hardening in the crystal and this is also known as region of easy glide. During this stage the dislocations are able to move relatively large distances without encountering any obstacle. Most of the dislocations in the first stage escape from the crystal surface and hence there is low strain hardening. Slip occurs on only one slip system at this stage.

Stage II : In second stage the curve is almost linear indicating that the strain hardening increases rapidly, slip occurs on more than one slip system. Many theories are put forward to, explain the increase in strain hardening during this stage.

Seeger assumed that strain hardening is due to pile up of dislocations. The pile up of dislocations on the slip planes at obstacles such as a sessile dislocation produces the back stress. The back stresses of the piled up dislocation block the movement of newly created dislocations and the back stress opposes the applied stress on the slip plane. Hence higher stress must be applied to break the obstacles or barriers. An important barrier that form in F.C.C. crystals is the Lomer-Cottrell barrier. Lomer-Cottrell barriers form in F.C.C. metals by slip on intersecting (111) plane.

According to Basinaki the strain hardening in Stage II is due to forest dislocations. In this theory it is assumed that dislocations moving in the slip plane cut through other dislocations intersecting the active slip pane. The dislocations passing through the active slip plane are called a dislocation forest. These forest dislocations arrest the movement of dislocations in the crystal and causes strain hardening.

Mott and Hirsch proposed a theory which is called as jog theory. According to them strain hardening depends on the jogs which hinder the motion of screw dislocations. Jogs in edge dislocation does not hinder motion. Jogs are produced either by moving dislocations cutting through forest dislocations or by forest dislocations cutting through source dislocations. These jogs are obstacles to the dislocation movement. At a high stress if the applied stress is increased these jogs will run into each other resulting local annihilation.

State III : In the third stage the rate of strain hardening decreases. In this stage the stresses are very high so that the dislocations struck up in stage II are able to move by a process that had been suppressed at lower stresses. Cross slip is the main process in Stage III by which the screw dislocations which were held up in stage II cross slip and possibly return to the primary slip plane by double cross slip. By this mechanism dislocations can over come the obstacles and move, hence stage III exhibits a low rate of work hardening.

3.4 RECOVERY, RECRYSTALLISATION AND GRAIN GROWTH

When a metal is cold worked, a finite fraction of the energy expended in cold work is stored in the metal as strain energy. And also cold working increases greatly the number of dislocations in a metal. A soft annealed metal usually have dislocation densities of the order of 10^6 to 10^8 dislocation lines per cm^2 and heavily cold worked metals will have approximately 10^{12} dislocation lines per cm^2. Therefore cold working has increased the number of dislocations in a metal by a factor as large as 10,000 to 1,000,000. As the dislocation density increases the strain energy of the metal increases since each dislocation is associated some amount of lattice strain. This stored strain energy produces internal stresses in a cold worked metal. In order to bring the cold worked metal back to its strain free condition by releasing the internal or residual stresses, a particular process of heat treatment called annealing is employed.

Similarly in wire drawing process during drawing the metal wire through a die, the metal gets work hardened and the resistance to further drawing increases thereby increasing the power. Finally the resistance to deformation becomes equal to the resistance to fracture and causes the wire to break upon further application of drawing stress. If at this stage the wire drawing process is stopped nothing will happen but if it is required to continue the drawing operation further to achieve the desired shape and size, the metal should be brought back to its original condition prior to deformation. This can again be achieved by a process called annealing.

Annealing involves heating the metal below its melting point so that the metal softens and returns back to a strain free condition and this can be achieved by the following three stages such as :

    (a) Recovery,

    (b) Recrystallisation, and

    (c) Grain Growth.

These stages and their general effects on properties and micro structures are shown schematically in Fig. 3.21 and 3.22.

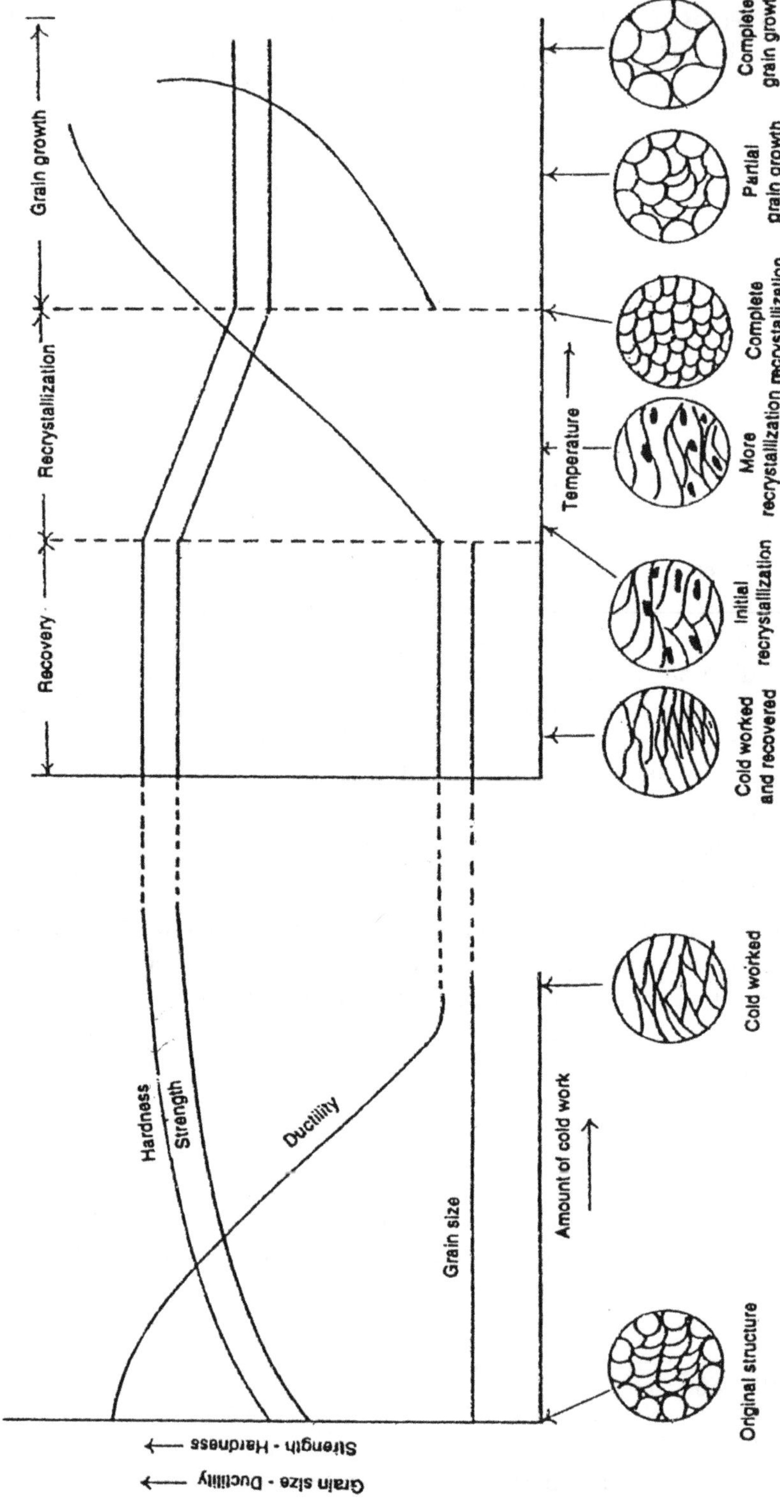

Fig. 3.21 Recovery, recrystallisation and grain growth.

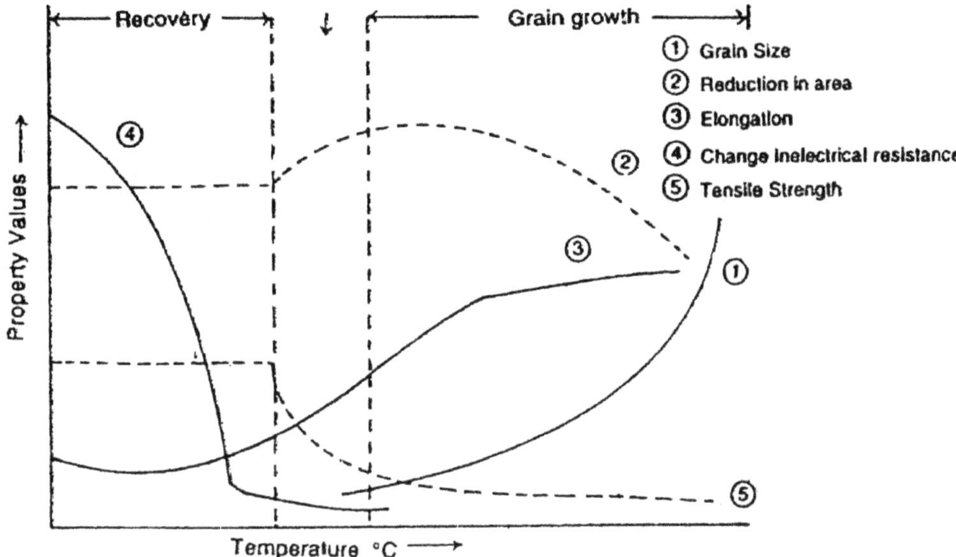

Fig. 3.22 *Effect recovery, recrystallisation and grain on material properties.*

3.4.1 Recovery

Recovery involves heating the metal to about 0.1 T_m , where T_m is the melting point of the metal in the absolute temperature scale. The main effect of recovery is to relieve internal stresses that are induced during cold working such as in rolled, drawn, and extruded objects. Recovery also prevents season cracking in brasses.

Recovery causes very little changes in mechanical properties and has no effect on microstructure. During recovery a rapid increase occur in electrical conductivity and only slight increase occur in tensile and yield strengths.

3.4.1.1 *Polygonisation*

During cold working the dislocations pile up at obstacles. And during recovery process these dislocations annihilate each other by migrating into lower energy planes, and rearrange themselves. This mechanism is known as polygonisation and the basic movement of dislocations during polygonisation is a climb process in which the dislocations climb out of their slip planes and rearrange themselves in a lower energy planes.

3.4.2 Recrystallization

Recrystallization occurs by a nucleation and growth process and it follows recovery. The recrystallization occurs at a higher temperature than recovery. During recrystallization period entirely new, strain free crystals are formed from the deformed metal and the distorted elongated grains disappear. The small strain free nuclei usually appear at the most severely deformed portions of the grains usually the grain boundaries and slip planes. Recrystallization can be detected by metallographic methods.

3.4.2.1 *Recrystallization Temperature*

Recrystallization temperature is the temperature at which a particular metal with a particular amount of cold deformation, will completely recrystallize in a definite period of time usually one hour. For pure metals the recrystallization temperature is about 0.3 T_m and for alloys it is 0.5 T_m where T_m is the melting point. Table below lists the recrystallization temperatures for different metals.

Metal	Recrystallization Temperature. °C
Nickel	600
Iron	450
Copper	200
Aluminium	150
Zinc	Room Temperature

Recrystallization temperature depends upon the following variables :

(a) *Amount of prior deformation* : If the degree of cold work is more then the recrystallization temperature will be lowered and grain size will also become smaller. If the degree of deformation is less then the temperature required to cause recrystallization will be high.

(b) *Particular metal* : Recrystallization temperature varies with the metal. It has one value for copper and another value for nickel.

(c) *Purity of Metal* : The recrystallization temperature decreases with increasing purity of the metal. Solid solution alloying additions always increases the recrystallization temperature. For example very pure aluminium (zone refined) crystallizes below room temperature while commercial aluminium recrystallizes at 150 – 200 °C.

(d) *Annealing Time* : Increased annealing time decreases the recrystallization temperature.

(e) *Initial Grain Size* : The finer the initial grain size of cold worked metal, the lower is the recrystallization.

3.4.2.2 *Effect of Recrystallization on Properties and Micro Structure*

(a) Recrystallization causes much change in mechanical properties. Strength and hardness decreases whereas ductility increases.

(b) Recrystallization causes the disappearance of distorted and elongated cold worked grains.

(c) Recrystallization does not change the crystal structure. B.C.C. metal remains as B.C.C.

(d) Recrystallization causes further relief of internal or residual stresses.

3.4.3 Grain Growth

Grain growth is the third stage of annealing that follows recrystallization. Recrystallization produces strain free small equiaxed grains. When the temperature is increased above that of recrystallization or holding for longer time after recrystallization, these strain free crystals grows in size by the coalescence of some of the new grains formed. Grain growth does not start from nuclei but growth occurs at the expense of their neighbouring grains as some grains being absorbed by others.

Strength and hardness decreases with grain growth but ductility increases.

3.4.3.1 *Factors Influencing Grain Growth*

The following factors influence the grain growth:

(a) *Annealing Temperature* : Grain growth increases with increase in annealing temperature.

(b) *Annealing time* : Grain growth is rapid during initial time period and then growth occurs more slowly with the passage of annealing time.

(c) *Rate of heating* : Slow heating causes the formation of few nuclei and favours grain growth.

(d) *Degree of prior deformation* : Severe deformation produces large number of nuclei during recrystallization and results in a small grain size whereas light deformation produces few nuclei during recrystallization and results in large grains size.

(e) *Insoluble impurities* : The greater the amount and the finer the distribution of insoluble impurities, the finer the final grain size.

(f) *Alloying elements* : Vanadium carbide particles restricts the grain growth of austenite during the heat treatment of high speed steel. Addition of alloying element like nickel restricts the grain growth during annealing in steel and other non-ferrous alloys.

Coarse grains are generally undesirable since they impairs mechanical properties and results in serious consequences if a metal consisting coarse grains is put in service.

3.5 HOT WORKING AND COLD WORKING

Mechanical working of a metal is nothing but performing plastic deformation to change its shape, properties and surface condition. Plastic deformation can be performed by two process such as

(a) Hot working

(b) Cold working

3.5.1 Hot Working

Plastic deformation of a metal above the recrystallization temperature but below the melting or burning point is called as hot working. Since high temperatures are involved, hot working requires lower stresses to produce plastic deformation and strain hardening does not occur at that high temperature because recrystallization occurs immediately after plastic deformation i.e., self annealing occurs. Hence hot working refines grain structure.

Hot working produces the same results as cold working and annealing together produces. Surface oxidation and scaling occurs during hot working and hence surface finish of hot worked metal is not as good as cold working.

The temperature at which to start hot work and at which to stop should be controlled carefully since this affects the properties of hot worked metal. Phase change may occur if the temperature is very low.

3.5.1.1 *Effects of Hot Working on Properties*

(a) Hot working improves elongation and reduction in area but directional properties are produced.

(b) Hot working does not introduce residual stresses

(c) Hot working does not affect tensile strength, hardness, corrosion resistance etc.

3.5.1.2 *Precautions in Hot Working*

Grain growth occurs whenever the working temperature is above recrystallization temperature hence to minimize large grains, the final working operation should be carried out below the recrystallization temperature.

3.5.1.3 *Limitations of Hot Working*

(a) Surface oxidation and scaling occurs during hot working since it is carried at a high temperature.

(b) Surface decarburisation also occurs.

3.5.1.4 *Hot Working Process*

Commonly employed hot working processes are:

(a) Forging

(b) Rolling

(c) Extrusion

(d) Drawing

(e) Piersing

3.5.2 Cold Working

Plastic deformation of a metal below its recrystallization temperature is called as cold working. Since low temperatures are involved, cold working requires higher stresses to produce plastic deformation and since strain hardening occur at this low temperatures, the stress increases with the amount of deformation. Hence annealing should be carried out to counter work hardening and to cause further plastic deformation. Cold working distorts grain structure.

Surface oxidation and scaling does not occur during cold working, since low temperatures are involved and hence good surface finish will be obtained.

Excessive cold working causes the formation and propagation of cracks in the metal. The loss of ductility during cold working impairs the machinability of a metal.

Dimensional accuracy is more in cold working. Cold working is more economical than hot working for ductile metals because no heating is required.

3.5.2.1 *Effect of Cold Working on Properties*

(a) Cold working increases yield strength, tensile strength, hardness, fatigue strength and decreases percentage elongation, impact strength, resistance to corrosion, reduction of area, and ductility.

(b) Cold working induces residual stresses.

3.5.2.2 *Precautions in Cold Working*

(a) Excessive cold working should be avoided since this leads to the formation and propagation of cracks.

(b) Residual stresses in the cold worked metal should be controlled.

3.5.2.3 *Limitations of Cold Working*

(a) During cold working strain hardening or work hardening occurs and hence the ductility of cold worked metal will be less.

(b) Cold working requires higher stresses than hot working for plastic deformation and hence more powerful and heavier equipments are needed for cold working.

3.5.2.4 *Cold Working Processes*

Commonly employed cold working processes are:

(a) Rolling

(b) Extrusion

(c) Deep drawing

(d) Shearing

(e) Squeezing

(f) Bending

3.6 MODEL QUESTIONS

1. Briefly write notes on various imperfections in metal crystals.

2. Explain the following:
   (i) Edge dislocation
   (ii) Screw dislocation

3. Define and explain Burgers Vector.

4. Explain the following:
   (i) Frank-Read source
   (ii) Climb of edge dislocation
   (iii) Jogs

5. Distinguish between elastic deformation and plastic deformation.

6. Explain the mechanism of plastic deformation by slip and twinning.

7. What is critical resolved shear stress? Calculate the critical resolved shear stress for slip.

8. Explain the deformation of polycrystalline materials.

9. Explain the principle of work hardening or strain hardening

10. Explain various stages of work hardening.

11. Differentiate between recovery, recrystallization and grain growth.

12. Explain the following:
    (i) Cold working
    (ii) Hot workings

4   MECHANICAL WORKING OF METALS

Mechanical working of a metal is to change dimensions (shape), properties or surface condition by deforming the metal plastically.

The various mechanical working processes are as follows:

    (i)   Forging

    (ii)   Rolling

    (iii) Extrusion

    (iv) Drawing of wires, rods and tubes

    (v)   Sheet metal working

4.1 FORGING

Forging is the oldest method of metal working, having its origin with the primitive black smith. During industrial revolution the arm of the smith is replaced by machinery and today there is a wide variety of forging machinery is available. At present forging produces parts ranging from a bolt to a turbine rotor.

Forging process involves the deforming of a hot metal piece to the desired shape by using compressive force. The force may be impact type like the blow from a hammer or a squeeze type as with a hydraulic press.

4.1.1 Forging Temperatures

For forging, the metal work piece must be heated to a proper temperature at which it will possess high plastic properties both at the beginning and at the end of the forging process.

Excessive temperatures may result in oxidation which causes wastage of metal.

Insufficient temperatures does not induce sufficient plasticity in the metal work piece so that it is difficult to shape the metal by hammering or pressing. Cold working defects such as hardening, cracking etc., are liable to occur if the forging temperature is insufficient.

The finishing temperature i.e., the temperature at which the hammering of a forging is left off, has an important influence on the properties of the forged part. Therefore, the finishing temperature should be such that at which no grain growth occur, so that the work piece possess a fine grain structure otherwise the mechanical properties will be badly impaired.

Table 4.1 gives the forging temperatures for different metals and alloys.

Metal or alloy	Forging temperature, $^{\circ}$C	
	Starting	Finishing
Mild Steel	1300	800
Wrought Iron	1275	900
Medium Carbon Steel	1250	750
High Carbon Steel	1150	825
Copper, Brass and Bronze	950	600
Aluminium and Magnesium alloys	500	350

Table 4.1 Forging Temperatures.

4.1.2 Forging Hammers

In forging hammers the force is supplied by a falling weight or ram and the deformation results from dissipating kinetic energy of the ram. The most important types of forging hammers are board hammer and steam hammer.

(a) Board Hammer

As shown in the Fig. 4.1, in the board hammer the upper die and ram are raised by friction rolls which are gripping the board. When the ram reaches the highest point of its stroke, the board is released, the ram drops under the influence of gravity to produce blow energy. Then the board is immediately raised again to produce another blow. Thus forging is done with repeated blows of 60 to 150 blows per minute depending on size and capacity.

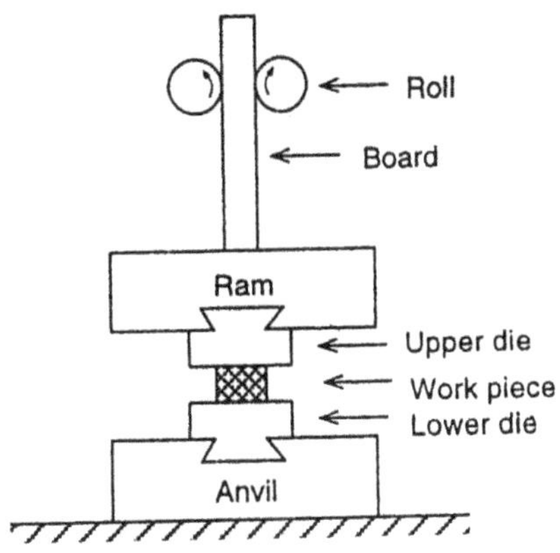

Fig. 4.1 *Board hammer.*

(b) Steam Hammer or Power Hammer

Compare with board hammer, more forging capacity is achieved with power hammer. In the power hammer, the ram is moved on the down stroke by steam or air pressure instead of gravity. Steam or air is also used to raise the ram on the upstroke. The air or steam enters the cylinder from one opening and forces the piston, piston rod and ram in the downward direction. Then the air or steam enters from the second opening and raises the ram vertically upward. Usually the air or steam pressure of about 6 to 8 kg/cm^2 is used. Power hammers are preferred for closed die forging and can produce forgings ranging in weight from a few kilograms to several tons.

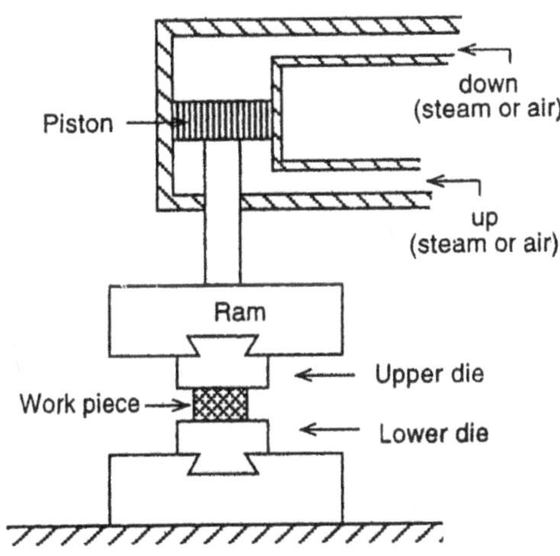

Fig. 4.2 *Steam or power hammer.*

In power hammer, the advantage over board hammer is that the energy of the blow can be controlled whereas in board hammer the mass and height of fall are fixed.

Forging hammers are the cheapest source of high forging load but the problems encountered are ground shock, noise and vibration.

4.1.3 Forging Presses

Forging presses are either mechanical type or hydraulic type. Presses produce forgings with accuracy compared with hammers. Presses apply pressure gradually instead of impact blows as in hammers.

(a) Mechanical Press

As shown in Fig. 4.3, most mechanical presses use an eccentric crank to transfer rotary motion into reciprocating linear motion of the press slide. Mechanical presses can operate at speeds varying from 35 to 100 strokes per minute with load ratings from 300 to 12000 tons. The die life is longer in mechanical press than in hammer since impact blows are not there and the pressure applied is of squeeze type. But the initial cost of the press is higher than hammer hence presses are used for mass production.

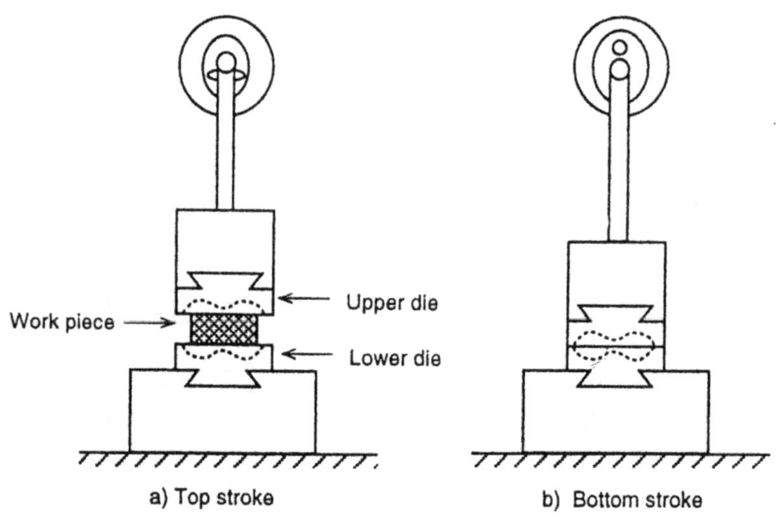

Fig. 4.3 Mechanical Forging Press.

(b) Hydraulic Press

Hydraulic press utilizes hydraulic pressure to move a piston in a cylinder either for downward stroke or for upward stroke. The working pressure of hydraulic fluid is upto 300 kg/cm^2. Hydraulic press is relatively a slow speed machine with load ratings from 500 to 18000 tons.

4.1.4 Forging Processes

The various forging processes are as follows :

    (A) Open Die forging or Smith forging

      (i)   Hand Forging

      (ii) Power Forging

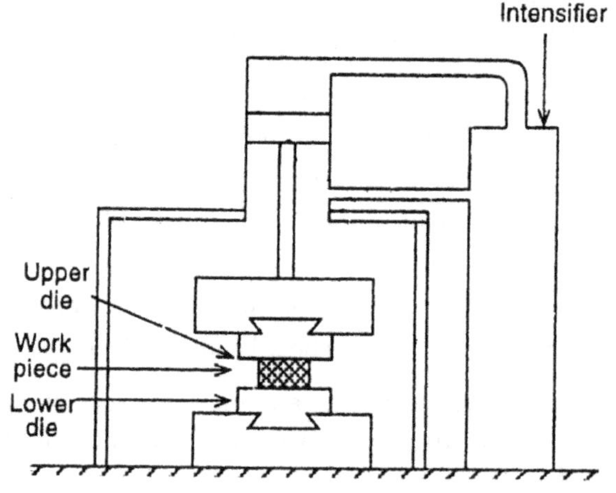

Fig. 4.4 Hydraulic Forging Press.

(B) Closed Die forging

   (i)   Drop Forging

   (ii)   Press Forging

   (iii)   Machine Forging or upset Forging

4.1.4.1 *Open Die Forging or Flat Die Forging*

Open die or flat die forging is also known as Smith forging. Open die forging is carried out between flat dies or dies of simple shape as shown in Figure 4.5.

Open die forging or Smith forging is of two Types

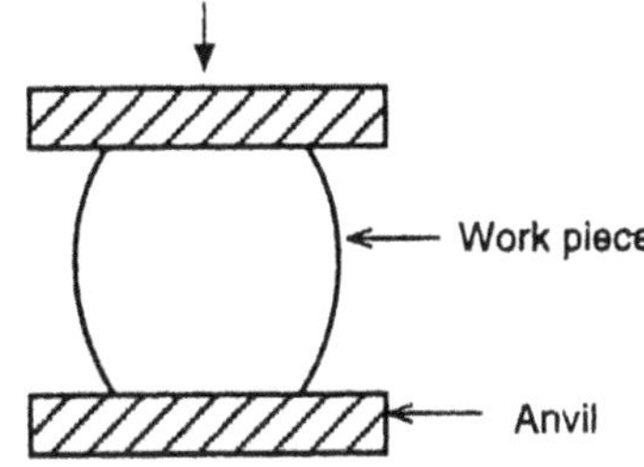

Fig. 4.5 Open die forging.

(a) Hand Forging : Smith forging when done by hand is known as hand forging. Hand forging is carried out by striking regular blows with suitable tool (a hammer) against the work piece which is placed on an anvil. Hand forging is employed to shape a small number of light forgings and in recent years hand forging is replaced by power forging.

(b) Power Forging : When Smith forging is carried out on a press or power hammer, it is known as power forging. In power forging, first the work pieces are heated to the desired temperature and then with the help of tongs the work pieces are brought under power hammer which delivers blows against the work pieces and deforms plastically by changing its shape to the desired shape.

Open die forging processes produce forged shapes of lesser accuracy than closed die forging since the work piece surface deform freely in open die forging. But open die forging requires simple tools and relatively inexpensive. Open die forging produces variety of shapes and sizes ranging from a few kilograms to about 250 tones.

4.1.4.2 *Drop Forging (or) Closed Die Forging*

To produce forgings in mass production drop forging is employed. Mass production justify the expensive dies. A closed die, a board hammer or a steam hammer (Figures 4.1 and 4.2) is used for making drop forgings.

The die for drop forgings consists of two halves. The upper half is fixed to the ram of the hammer and the lower half is fixed to the anvil. The stock of work piece which is heated to the plastic condition is placed in the lower die while the ram delivers the blows.

Drop forging may be carried in either single or multiple impression dies. The single impression type is suitable for forgings of simple shape and size such as gear blanks. The multiple impression type is suitable for forgings of irregular shape such as connecting rod, lever, etc.

In case of multiple impression dies the following operations are carried out :

The first operation is to reduce the size of the bar stock to the desired dimensions. This operation is called fullering impression.

The second operation is to form the bar stock to rough shape. This operation is called blocking impression or semi finishing impression.

The third operation is called final impression or finishing impression where the actual shape required is obtained.

The fourth operation is called trimming. Usually, in order to ensure that the metal fills the die cavity completely, it is customary to use a slight excess of metal than required. In the finishing operation when the dies come together, the excess metal squirts out of the cavity as a thin ribbon of metal which is called 'flash'. In the trimming operation the extra flash present around the forging is trimmed with a trimming die.

The parts obtained by drop forging are crank, crank shaft, connecting rod, crane hooks etc.

4.1.4.3 *Press Forging*

Press forging is similar to drop forging except that the metal is shaped to the desired shape by means of a single continuous squeezing action obtained by mechanical or hydraulic presses (Figures 4.3 and 4.4) and not by means of several blows as in drop forging. This squeezing action reduces noise and vibrations.

The impressions obtained in press forgings are clear compared to that of drop forgings where   jarred impressions are likely obtained. Presses gives a faster rate of production because the die is filled in a single stroke of the press.

The operations involved in press forging is similar to drop forging such as fullering, blocking, finishing and trimming.

Mechanical presses can operate at speeds varying from 35 to 100 strokes per minute with load ratings from 300 to 12000 tons.

Hydraulic presses are relatively slow speed machines with load ratings from 500 to18000 tons.

4.1.4.4 *Machine Forging or Upset Forging*

Even though drop and press forging are also done by machines, historically upset forging is called as machine forging. This involves an upsetting operation

Usually the upsetting is done in a horizontal machine. Fig. 4.6 shows the sequence operations in upset forging. First die heated bar stock is inserted between movable and fixed halves of the set of dies upto stop. Next the movable die grips the bar stock so that a recess is formed in the closed dies for shaping the projected stock. Stop is then brought to its idle position. Then the punch advances to upset the bar end and produces the finished forging.

Upsetting forging is used for making shafts, gear blanks, axles, bolts, nuts etc.

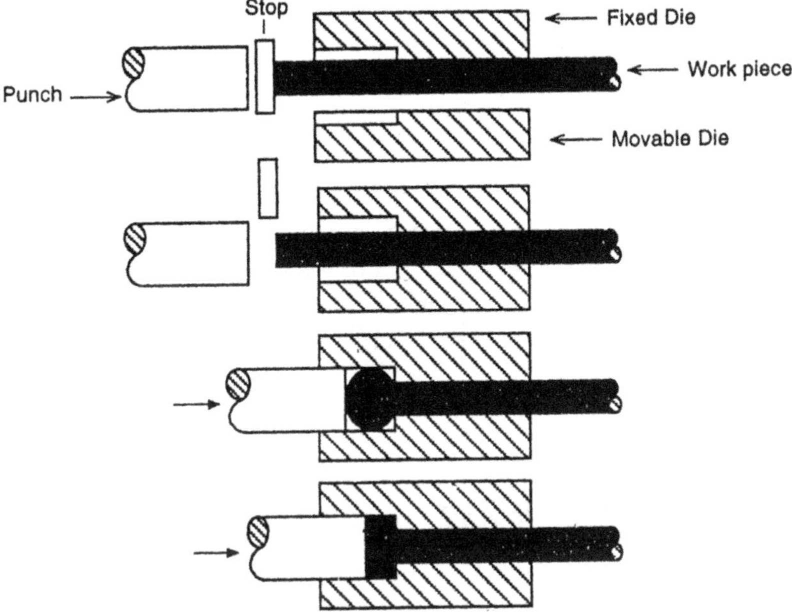

Fig. 4.6 Upset Forging Sequence.

4.1.5 Forging Defects

The most commonly found forging defects are as follows:

    (a) Defects resulting from defective ingots such as pipe, seams and segregation.

    (b) Defects resulting due to improper heating such as dirt, slag and blow holes.

    (c) Defects resulting from wrong forging methods such as seams, cracks, laps etc.

    (d) Defects resulting from incorrect forging conditions i.e., excessively high or low final forging temperatures such as burnt metal, decarburised steel etc.

    (e) Defects resulting from uneven cooling of the forged part such as residual stresses.

By chipping, shallow cracks and cavities can be eliminated. By grinding surface cracks and decarburised areas are removed. By using presses distorted forgings are straightened. By heat treatment such as annealing or normalizing, residual stresses can be eliminated.

4.2 ROLLING

Rolling is the process of plastically deforming metal so that shaping of metal into the semi finished or finished condition is obtained by passing the metal between rolls. Rolling is the most widely used metal working process since it gives high productivity at low cost. Various shapes such as plate, sheet, rod, bar, pipe and sections such T,L,I and channel can be produced by rolling.

During rolling the metal between rolls is subjected to high compressive stresses from the squeezing action of rolls and also subjected to surface shear stresses as a result of the friction between the rolls and the metal. For drawing the metal into the rolls, the frictional forces are also responsible.

4.2.1 Hot Rolling and Cold Rolling

In hot rolling, the metal is heated to above recrystallization temperature and then fed between rolls. This leads to grain refinement, i.e, fine grain structure is obtained in hot rolled parts thereby improve mechanical properties. During hot rolling, work hardening does not occur and coefficient of friction between the rolls and the metal is higher. Hot rolled parts does not have good surface finish since oxidation of the surfaces occur at high temperature.

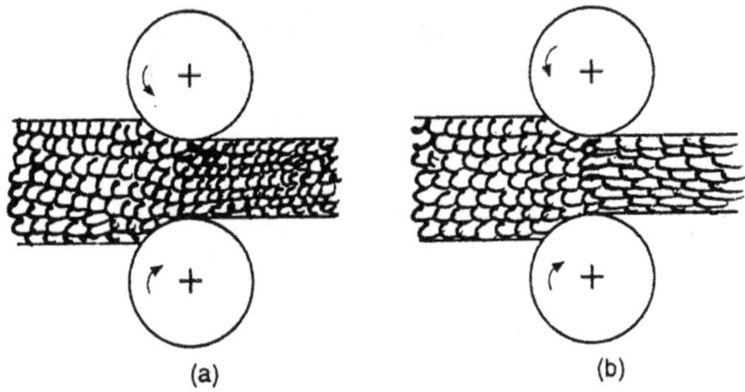

Fig. 4.7 (a) Hot rolling (b) Cold rolling.

In cold rolling, the metal is fed between rolls below recrystallization temperature. This results in elongated grain structure. During cold rolling, work hardening occur and thereby increases hardness and decreases the ductility of metal. The coefficient of friction between the rolls and the metal is lower. Cold rolled parts have smooth surface finish since oxidation does not occur thereby the surfaces are oxide free.

4.2.2 Types of Rolling Mills

According to the number and arrangement of rolls, rolling mills are classified as follows:

    (a) Two - high rolling mill

    (b) Three - high rolling mill

    (c) Four - high rolling mill

    (d) Tandem rolling mill

    (e) Cluster rolling mill

4.2.2.1 *Two - High Rolling Mill*

Two-high rolling mills are of two types :
Pull over and Reversing (Fig. 4.8).

In pull over mill, rolls of equal size are
rotated only in one direction and the rolled
metal is returned to the entrance or rear of
the rolls for further reduction by hand or by
means of a platform which can be raised to
pass the work above the rolls.

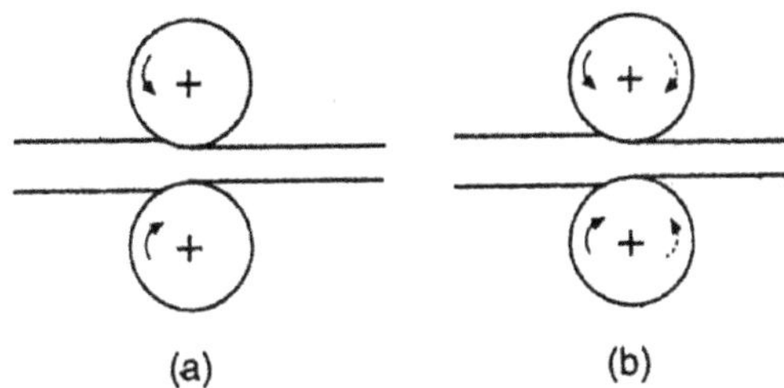

Fig. 4.8 Two – high rolling mill (a) Pull over (b) Reversing.

In reversing mill, first the rolls rotate in
one direction and then in the other direction so that the rolled metal may be passed back and
forth through the rolls many times thereby improving the speed of operation.

4.2.2.2 *Three - High Rolling Mill*

Three-high rolling mill consists of three rolls mounted one
above the other. In this arrangement the top and bottom
rolls rotate in the same direction and the middle one in
opposite direction (Fig. 4.9). That means the adjacent rolls
rotate in opposite directions so that the material may be
passed between the top and middle rolls in one direction
and the bottom and middle rolls in the opposite direction.
This arrangement eliminates the need of reversing the rolls.
First the material passes through the bottom and middle
rolls and then returning to pass through the top and middle
rolls so that the size is reduced at each pass. Much less
power is required for three -high rolling mill.

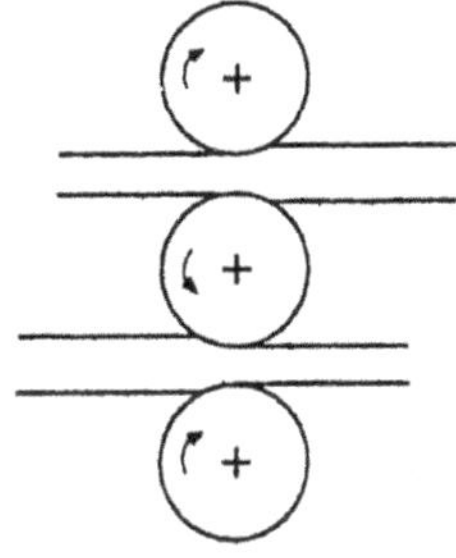

Fig. 4.9 Three - high
rolling mill.

4.2.2.3 *Four - High Rolling Mill*

Four-high rolling mill consists of four parallel rolls one
above the other in which two smaller in size and two
bigger in size (Fig. 4.10). The bigger rolls are called back
up rolls because they reinforce the smaller rolls to
minimise roll deflection thereby minimising the tendency
of producing plates and sheets thicker at the centre than at
the two outer edges. The two smaller rolls are called work
rolls in which the work piece is fed. The bigger rolls rotate
in opposite direction and the smaller rolls also does the
same. Four-high rolling mill is commonly employed for
both hot and cold rolling of plates and sheets.

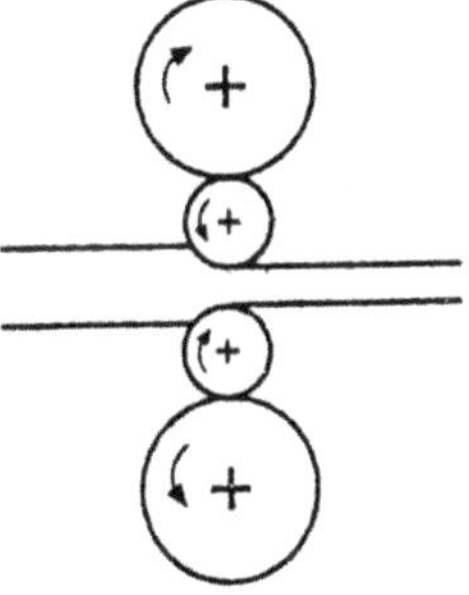

Fig. 4.10 Four - high
rolling mill.

4.2.2.4 Tandem Rolling Mill

Tandem rolling mill consists of several stands of four high rolling mills arranged one after the other so that as the material comes out of one set of rolls, it enters the second, third, fourth and so on and finally comes out in required size and shape (Fig. 4.11).

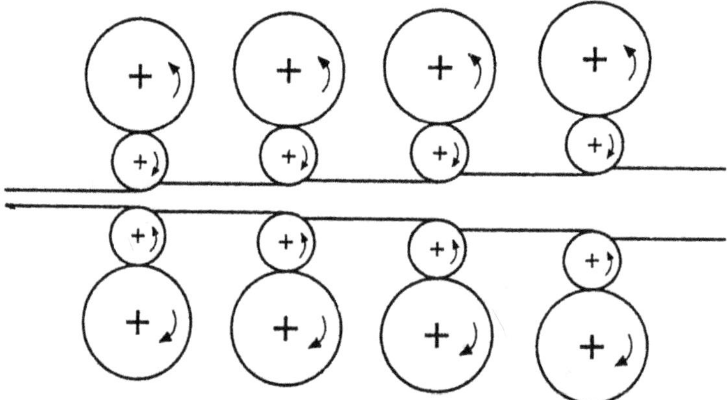

Fig. 4.11 *Tandem rolling mill.*

This arrangement works out to be conomical and mass production is possible.

4.2.2.5 Cluster Rolling Mill

In cluster mill, each of the work rolls is supported by two backing rolls (Fig. 4.12). With this arrangement very thin sheet can be rolled to very close tolerances.

4.2.2.6 Stages in Rolling

Fig. 4.13 shows how a 4" × 4" billet is reduced to a 1/2" diameter round bar in various passes with gradually changing geometry gap between two rolls.

4.2.2.7 Roll Piercing (tube making)

Roll piercing is employed to produce seamless tubes.

Seamless tubes are made from solid round billets. In this process, the solid round billets are heated above recrystallization temperature and then

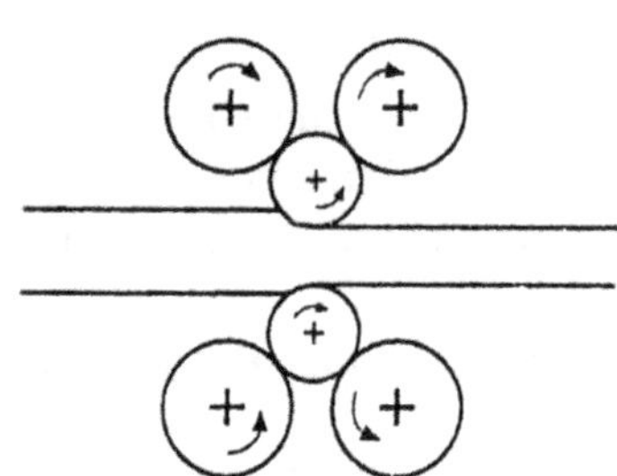

Fig. 4.12 *Cluster rolling mill.*

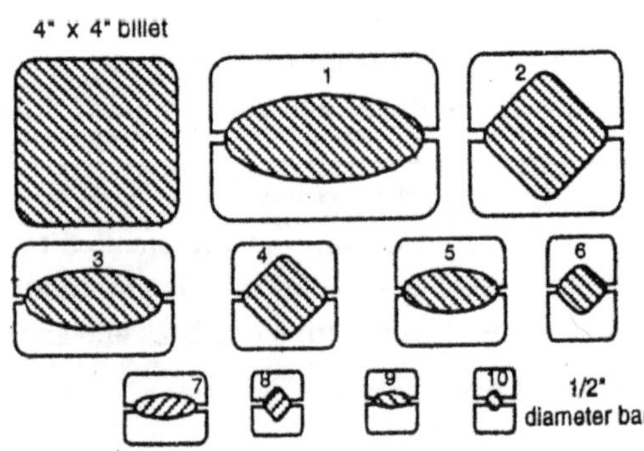

Fig. 4.13 *Stages in rolling a round bar.*

passed between two conical rolls which impart axial as well as rolling movement to the billet and force it over the pointed mandrel. The pointed mandrel pierces the billet to open up a seam in its center (Fig. 4.14). This produces a thick walled tube which is then subsequently passed over mandrels between grooved rolls such as plug rolling mill, reeler rolls, sizing rolls in order to obtain the required diameter, thickness, accuracy and surface finish for the tubes.

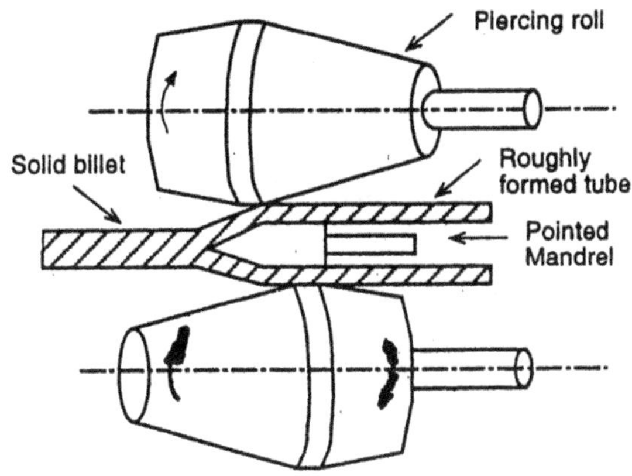

Fig. 4.14 *Roll Piercing.*

4.3   EXTRUSION

Extrusion is the process in which a block of metal is reduced in cross section by forcing it to flow through a die orifice under high pressure. The high pressure is obtained by hydraulic presses or mechanical presses. The extrusion dies are made with heat resistant steel or tungsten carbide which can withstand the high extrusion pressures.

Usually extrusion process is used to produce cylindrical bars or hollow tubes but irregular shapes can also be produced.

The two basic types of extrusion are :

   (a)   Direct extrusion or Forward extrusion

   (b)   Indirect extrusion or Backward extrusion

4.3.1 Direct Extrusion

As shown in the Fig. 4.15, in this process the metal billet is kept in the container and the required force is applied by the ram so that the meal billet is driven through the die. The ram compresses the metal billet against the container walls and the die plate, thus forcing it to flow through the die opening called orifice, thereby acquiring the shape of the orifice. A dummy block or pressure plate is kept between the hot billet and the ram so that it protects the ram from the heat and pressure.

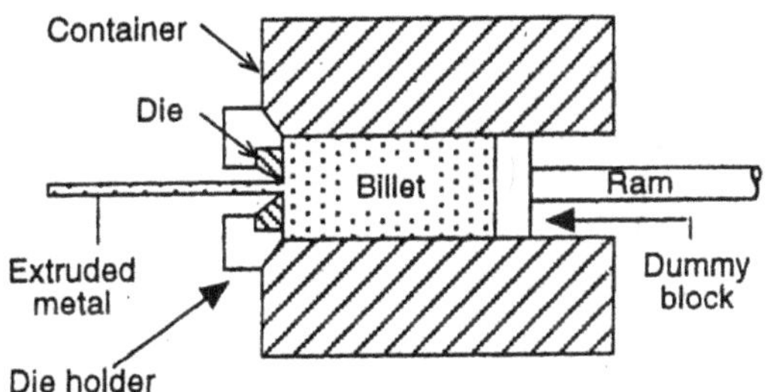

Fig. 4.15 *Direct Extrusion Process.*

In this process, the direction in which the extruded metal leaves the die is the same as that of ram motion and hence the name forward extrusion.

In forward extrusion, due to the relative motion between the container walls and the metal billet, friction occurs. To combat this friction, lubricants such as oil, graphite and sometimes molten glass are employed.

4.3.2 Indirect Extrusion

As shown in Fig. 4.16, in this process, the ram which carries the die, compresses the metal billet against the container closure plate, thus forcing it to flow through the die orifice thereby acquiring the shape of the orifice.

In this process, the extruded metal leaves the die in the direction opposite to the ram motion and hence the name backward extrusion.

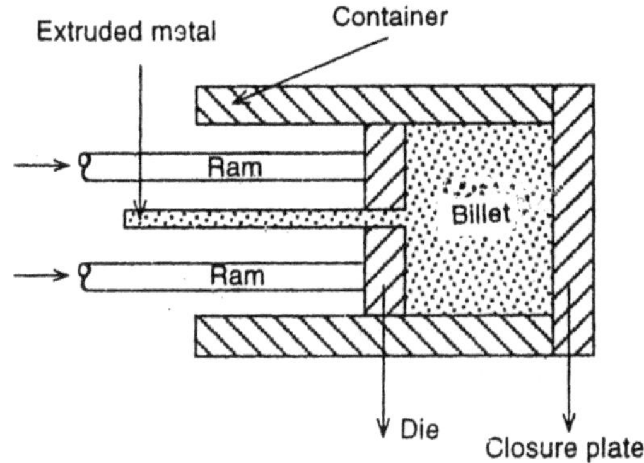

Fig. 4.16 Indirect Extrusion process.

In backward extrusion the metal billet in the container remains stationary and hence there is no relative motion between container walls and the metal billet, the friction forces are lower and the power required is less than forward extrusion.

4.3.3 Impact Extrusion

As shown in Fig. 4.17, in this process the metal blank i.e., work piece is placed in the die and the punch strikes the metal blank against the die. Then the metal is extruded through the gap between the punch and die, opposite to the punch motion. The extruding force is usually obtained by a mechanical press. Impact extrusion is used for softer metals such as aluminium, copper, lead and tin. Collapsible medicine tubes and tooth paste tubes are made by impact extrusion.

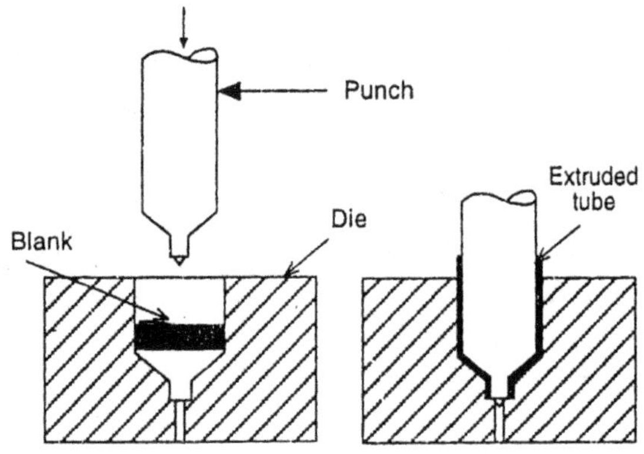

Fig. 4.17 Impact Extrusion Process.

4.3.4 Hot Extrusion and Cold Extrusion

Hot extrusion process include direct extrusion and indirect extrusion. In this process, heated billet is used as a starting material. Extrusion pressure are lower. For hot extrusion of steel, the billets are heated in the range 1100 to 1200 °C and extrusion pressures usually range from 1,25,000 to 1,80,000 psi.

Cold extrusion processes include direct extrusion, indirect extrusion and impact extrusion. In the cold extrusion the billet is not heated and the extrusion pressures required are higher. Cold extrusion is usually employed for producing simple shapes with better surface finish and mechanical properties.

4.4 DRAWING OF WIRES, RODS AND TUBES

Drawing operation is carried out by pulling the metal through the die by means of a tensile force which is applied to the exit side of the die. But most of the plastic deformation occur due to compressive force between the metal and the die. The reduction in diameter of a solid bar or rod by successive drawing is called as bar, rod or wire drawing depending on the diameter of the final product. When a hollow tube is drawn through a die with the support of mandrel, the process is called tube drawing. Rod, wire and tube drawing is usually carried out at room temperature.

4.4.1 Wire Drawing

Wire drawing is carried out by pulling a rod through the die which causes reduction in cross sectional area of the rod (Fig.4.18). The tensile force used for pulling must not exceed the yield stress of the metal being drawn. The size reduction is obtained by the compressive forces that are induced by tensile pull. These compressive forces must be above the yield stress of the metal being drawn.

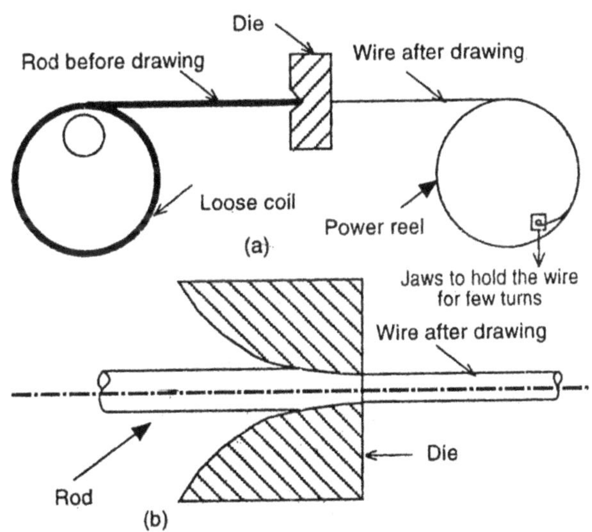

Fig. 4.18 *(a) Wire drawing equipment (b) Sectional view of wire drawing.*

The raw material for wire drawing is hot rolled wire rod. The rods obtained from hot rolling are first cleaned with acid baths to remove scale and rust and then coated with lubricants such as soaps or grease. The two commonly used methods for wire drawings are single draft and continuous wire drawing.

In single draft method, the wire is drawn through one die and for further reduction in size many similar dies will be required. In this method the drawing operation is started by making a pointed end in the rod and pushing it through the tapered hole in the die. The pointed end is gripped by tongs and sufficient wire is pulled through the die and then attached to a power operated reel. The reel is then rotated at a calculated speed so that the wire will be drawn through the die at a desired rate. During operation, to minimise friction, the area of contact between the die and the wire is lubricated continuously.

In continuous drawing, the wire is drawn continuously through a number of dies which are arranged in series. The number of dies depends upon the reduction required and also on the kind of metal being drawn. Depending on the metal and the reductions involved, intermediate annealing should be carried out in order to remove work hardening effects which are caused due to cold working.

Wire drawing dies are usually made up of chilled cast iron, hardened alloy steets, tungsten carbide or diamond. But of these, tungsten carbide is more popular die material for wire drawing since it possess high wear resistance.

4.4.2 Rod Drawing

The principle involved in rod drawing is same as that of wire drawing. The rod which can not be coiled are produced on draw benches. First the end of the rod is pointed to enable it to be passed through the hole of the die. After inserting through the die, the rod end is then gripped and attached to the jaws of the draw bench which pulls the rod through the die, thereby reducing its diameter and increasing its length (Fig. 4.19). Draw bench is moved either by a chain drive or by a hydraulic mechanism.

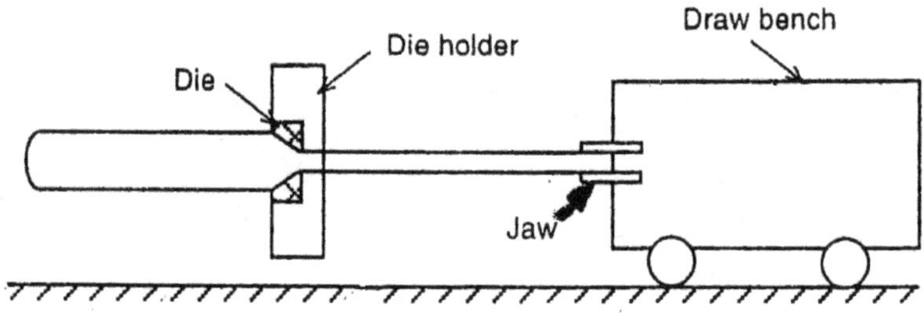

Fig. 4.19 Rod Drawing.

4.4.3 Tube Drawing

Tube drawing is also similar to other drawing processes. Hollow tubes which are produced by hot forming processes such as extrusion or piercing and rolling are often finished to close dimensional tolerances with better surface finish by cold drawing.

Cold drawing reduces the diameters of the tubes, produces tubes of irregular shape and increases the mechanical properties of the tubes by work hardening.

The three basic types of tube drawing processes such as sinking, fixed plug and moving mandrel are shown in Fig. 4.20.

In tube sinking, the inside of the tube is not supported thereby the wall thickens slightly and the internal surface becomes uneven.

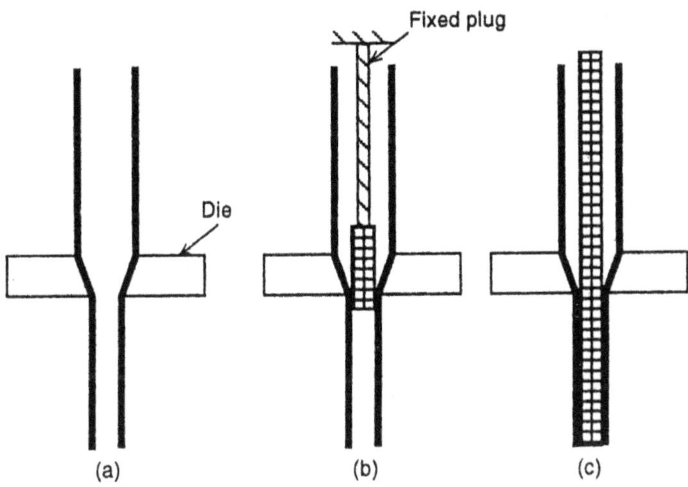

Fig. 4. 20 *Methods of tube drawing (a) Sinking (b) Fixed plug (c) Moving mandrel.*

In fixed plug drawing, the cylindrical, or conical plug controls the size and shape of the inside diameter and thereby produces tubes of greater dimensional accuracy than obtained in tube sinkimg. The reduction in area rarely exceeds 30 percent due to increased friction from the plug.

Tube drawing with a long mandrel (a long hard rod) the frictional problems are minimised.

4.5 SHEET METAL WORKING

The most important sheet metal working processes are as follows:

    (a) Blanking

    (b) Piercing

    (c) Coining

    (d) Deep Drawing

    (e) Spinning

    (f) Embossing

4.5.1 Blanking

Blanking is the shearing operation which shears the metal to closed contours. The removed portion which is detached from the metal strip is called as blank and is used for further operations. The remaining metal goes to the scrap.

The basic tools required for blanking along with metal working press are punch and die (Fig. 4.21). The blanking operation is performed at one stroke of the press which makes the punch to penetrate into the sheet metal and produces the required blank.

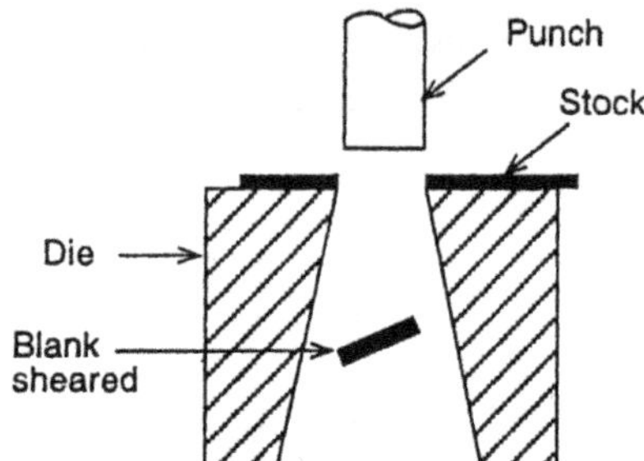

Fig. 4.21 Blanking operation.

4.5.2 Piercing

Piercing also called sometimes as punching, is used for making holes in a sheet. Piercing is similar to blanking, except that the punched out portion coming out of the die is discharged as scrap in piercing. In piercing, a pointed punch is forced through the sheet metal to obtain a hole.

4.5.3 Coining

Coining is a closed die squeezing operation except that the flow of the metal occurs only at the top layers and not the entire volume.

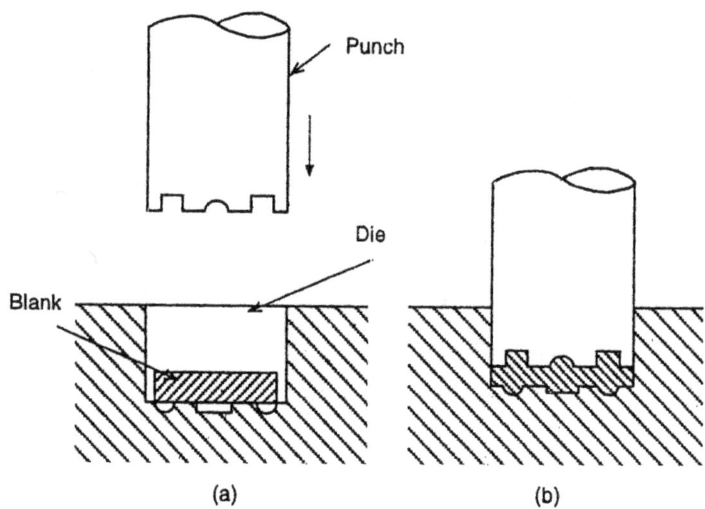

Fig. 4. 22 Coining operation
(a) Punch and die (b) Penetration of the punch.

Fig. 4.22 shows the coining die and punch on which the desired design required on both sides of the final object is engraved. First the blank is placed in the die and then the punch descends on to the blank under high pressure (of the order of 1600 Mpa). Under this high pressure the metal flows into the crevices of the engraved design. Very high pressure is required for coining in order to reproduce very fine details of the design.

Coining is employed for producing coins, medals and for designs on decorative pieces.

4.5.4 Deep Drawing

Deep drawmg process is used for shaping flat blank into a cup shaped components without any appreciable change in sheet thickness.

Deep drawing is carried out by placing a blank of proper size over a shaped die and pressing the blank into the die with a punch under pressure which is obtained either by a hydraulic press or by a mechanical press (Fig. 4.23). During the operation, to prevent wrinkling a blank holder is used.

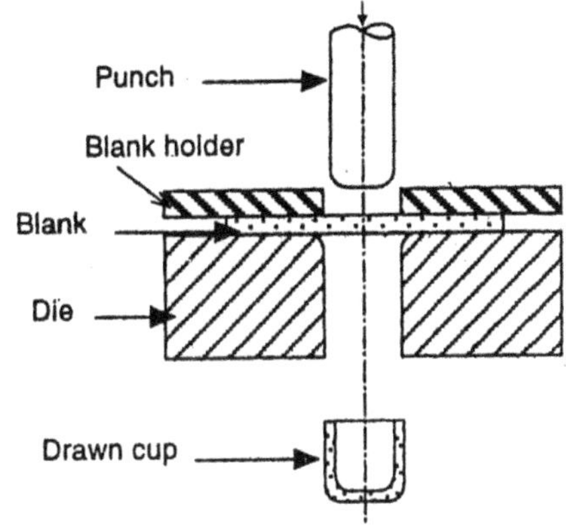

Fig. 4. 23 Deep drawing operation.

4.5.5 Spinning

Spinning is the process employed for producing cup shaped articles and other deep parts of circular symmetry. In spinning, the blank which is fixed against the form block is rotated and then a gradually moving force is applied on the blank so that the blank takes the shape of the form block..

Fig. 4.24 shows the typical spinning operation set up. The spinning machine is similar to ordinary lathe. The form block which has the shape of desired object is fixed to the head stock of the spinning machine. The blank is held against the form block with the help of tail stock. After clamping. the blank is rotated at its operating speed. The blank is progressively formed against the form block by pressing with hand tool so that it conforms to the shape of the form block.

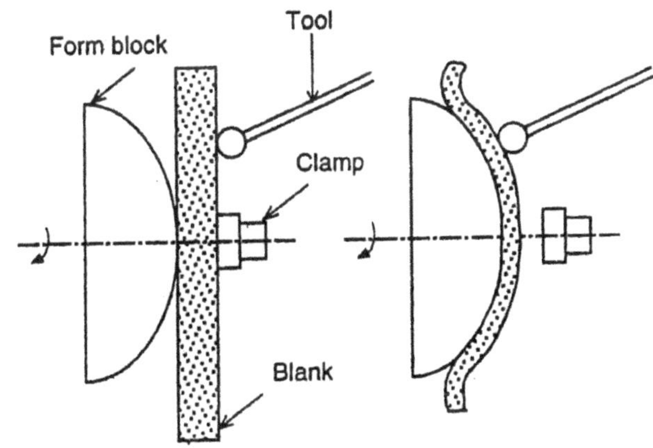

Fig. 4. 24 Spinning operation.

The advantage of spinning process is that it is very cheap process since the equipment cost and tool cost is low. Some complex parts can be economically produced by spinning.

The limitation of the process is that the process depends very much on the operators skill to fold the metal over form block..

4.5.6 Embossing

Embossing operation is used in making raised figures and shallow indentation on sheets with its corresponding relief on the other side.

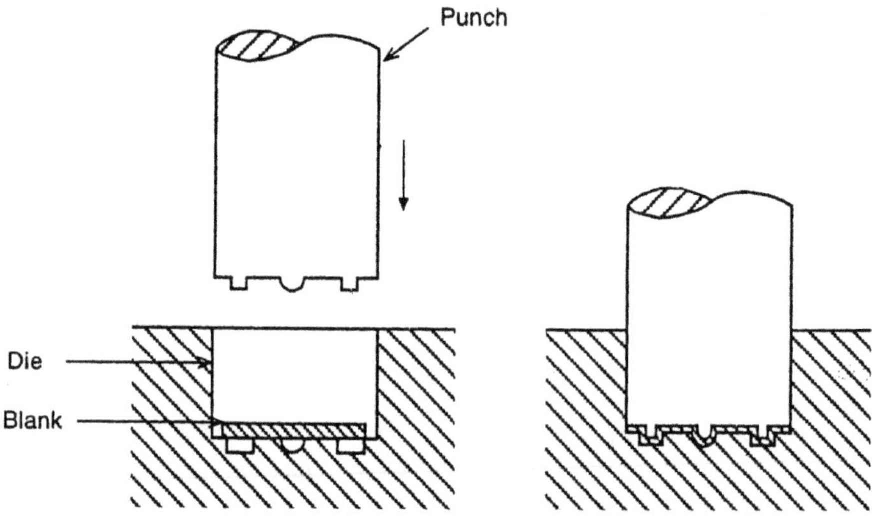

Fig. 4. 25 *Embossing operation.*

Fig. 4.25 shows a typical embossing operation. The embossing equipment consists of a punch and die with required contours, so that the required design is obtained when the punch presses the blank which is kept in the die. Embossing slightly differs from coining. In embossing the thin blank will have the same design on both sides, one side being depressed and the other is raised whereas coining produces different designs on each side.

4.6 MODEL QUESTIONS

1.   What is forging ? List out the forging temperatures for various metals.

2.   With neat sketches explain the following:
     (i)   Board hammer
     (ii) Steam hammer

3.   Explain the principle involved in mechanical press and hydraulic press.

4.   Explain the following:
     (i)   Open die forging
     (ii) Drop forging

5.   Explain the sequence of operations involved in upset forging.

6. Distinguish between hot rolling and cold rolling. Explain the principle involved in rolling process

7. With neat sketches explain the following:
    (i)   Two - high rolling mill
    (ii) Four - high rolling mill
    (iii) Tandem rolling mill

8. With neat diagram explain the process of tube making by roll piercing.

9. Differentiate between forward extrusion and backward extrusion.

10. Describe the forward extrusion process with neat diagram.

11. Write a note on impact extrusion process.

12. Explain the following with neat sketches:
    (i) Wire drawing
    (ii) Tube drawing

13. Write short notes on the following:
    (i)   Blanking
    (ii) Deep drawing
    (iii) Spinning

14. Differentiate between coining and embossing.

15. With neat diagram explain the process of spinning.

5 FOUNDRY

Foundry is one of the oldest industry in the world. Foundry industry is an establishment which produces castings commercially. The metallic object obtained by pouring the molten metal into a mould (the shape of the mould cavity is that of the required casting) and allowing it to solidify is known as casting. Almost all metals/alloys can be cast practically.

Castings find wide applications in modern equipment used for communication, power, transportation, agriculture, construction and for making various machines and tools.

The basic steps involved in producing a casting are as follows:

    (i)   Pattern making

    (ii)   Mould and core making

    (iii)   Melting of metals

    (iv)   Pouring molten metal

    (v)   Cleaning of the casting

    (vi)   Testing and inspection of the castings

5.1 PATTERN MAKING

Preparing a pattern is known as pattern making which is the first step in the casting process.

A pattern is a model (having its shape and size similar to the casting to be produced) which is prepared with a suitable material and is used for making the mould cavity in a suitable moulding material. When this mould cavity or mould is filled with molten metal and allowed it to solidify, it forms a reproduction of the pattern which is known as casting.

5.1.1 Pattern Materials

The selection of a particular material for pattern making depends on the following factors :

    (i)   Number of castings to be produced and the possibility of repeat orders

    (ii)   Type of moulding methods and equipment used

    (iii)   Degree of accuracy and surface finish required

    (iv)   Possibility of design changes

The following materials are generally used for making patterns :

    (i)   Wood

    (ii)   Metals and alloys

    (iii)   Plasters

    (iv)   Plastics

    (v)   Waxes

Wood : Wood is the most common material used for making the patterns. The main reasons for its popularity are :

    (i)   its cheapness

    (ii)   easily available

    (iii)   easily shaped, worked and joined to form any complex shape

    (iv)   light in weight

    (v)   its surface can be made smooth very easily by sanding

    (vi)   its surface can be preserved by applying a coating of shellac.

The most common draw backs of wood are :

    (i)   its susceptibility to moisture thereby it swells or shrink

    (ii)   low resistance to wear thereby it wears out quickly as a result of sand abrasion

    (iii)   its poor strength thereby it tends to break on rough usage

Due to the above reasons wooden patterns are generally used when a small number of castings are to be produced.

Before wood is used for pattern making it should be thoroughly dried and it should be straight grained, free from knots and free from excessive sap wood.

The woods generally used for pattern work are :

    (i)   Pine wood

    (ii)   Teak wood

    (iii) Mahogany

    (iv) Walnut

    (v)   Deodar

Metals and alloys : Metallic patterns are advantageously used when repetitive production of castings is required in large quantities.

Advantages of metallic patterns are :

    (i)   They are not affected by moisture

    (ii)   Less wear thereby they can withstand abrasion and rough handling

    (iii) High strength

    (iv) Easier to get precision and intricate shapes.

Disadvantages of metallic pattern are :

    (i)   Expensive than wood

    (ii)   Higher weight

    (iii) Tendency to get rusted

The metals commonly used for making patterns are :

    (i)   aluminium alloys

    (ii)   cast Iron

    (iii) steel

    (iv) copper base alloys

Plasters : Plaster patterns are made with gypsum cement (plaster of paris) which has a high compressive strength (upto 300 kg/cm^2) and controlled expansion (i.e., it expands on solidificalion). Hence by selecting a plaster of proper expansion rate it is possible to compensate the shrinkage of the casting so that shrinkage allowance need not be provided on the pattern.

Proprietary varieties of gypsum plasters such as Ultracal, Hydrocal, Hydrostone are also available with different setting times, expansion rates and compressive strengths to suit the varying requirements of pattern maker.

Plastics : Plastics (phenolic thermosetting resins) are now finding their way into foundries as a modern pattern material.

Advantages of plastics are :

    (i)   They do not absorb moisture

    (ii)   Light Weight

    (iii) They are strong

    (iv) High resistance to wear

    (v)   High resistance to chemicals

    (vi) Very smooth and glassy surface so that the pattern can be withdrawn from the mould easily.

To produce plastic patterns, first a mould in two halves is prepared with plaster of paris. Then the thermosetting resin is poured into the mould and the mould is subjected to heat to a specified temperature. The resin then solidifies and gives plastic pattern.

Waxes : Wax patterns are used in investment casting process. They are usually used as blends of several types of waxes and other additives. The most commonly used waxes are paraffin wax, carnauba wax, shellac wax, bees wax, cerasin wax. The wax patterns imparts a high degree of surface finish and dimensional accuracy to castings.

5.1.2 Pattern Allowances

Pattern allowances are very important in the design of a pattern. Although the pattern is used to produce a casting of required dimensions, it is never kept dimensionally identical with that of casting. This is because during solidification and after that the casting is subjected to various metallurgical and mechanical effects and hence to compensate these effects the following allowances must be made on the pattern if the final casting is to be of required dimensions.

    (i)   Shrinkage or contraction allowance

    (ii)   Machine finish allowance

    (iii) Draft or taper allowance

    (iv) Rapping or shake allowance

    (v)   Distortion or camber allowance

5.1.2.1 *Shrinkage Allowance*

Almost all cast metals shrink or contract volumetrically when cooled from molten state due to inter atomic vibrations. Hence to compensate this shrinkage and to obtain the final casting of required size, the pattern is made larger by an amount equal to that of shrinkage or contraction. Typical shrinkage allowances for important cast metals are given in table 5.1

	Contraction (percent)	Contraction (mm per metre)
Grey cast iron	0.7 to 1.05	7 to 10.5
White cast iron Metal	2.1	21
Malleable iron	1.5	15
Steel	2.0	20
Brass	1.4	14
Aluminium	1.8	18
Aluminium Alloys	1.3 to 1.6	13 to 16
Bronze	1.05 to 2.1	10.5 to 21
Magnesium	1.8	18
Zinc	2.4	24
Manganese steel	2.6	26

Table 5.1 *Shrinkage allowances for important cast metals.*

Table 5.1 shows that different metals will shrink at different rates because it is the property of the particular cast metal or alloy. Table 5.1 is only a guide line because the actual shrinkage of cast metal depends on the following factors.

(i) Composition of metal

(ii) Method of moulding

(iii) Moulding material and its hardness

(iv) Resistance of mould to shrinkage

(v) Design and intricacy of the casting

(vi) Pouring temperature of the molten metal

5.1.2.2 *Machine Finish Allowance*

Machine finish allowance should be given on pattern in order to compensate the amount of metal that is lost in machining or finishing the casting to obtain the final casting of required dimensions and surface finish. The casting will become undersize after machining if this allowance is not given on the pattern. Hence extra metal is to be provided which is to be subsequently removed by machining or finishing process.

The amount of this allowance depends upon the size of casting, type of machining operation such as grinding, turning, milling, boring etc., type of moulding process such as sand casting, die casting etc., and the degree of surface finish. Table 5.2 gives approximate machining allowances for various metals.

Metal	Machining allowance (mm)
Cast iron	
(i) Medium Castings	3
(ii) Large castings	10
Steel	
(i) Medium castings	4.5
(ii) Large Castings	12
Non ferrous metals (Brass, Bronze, Aluminium etc.,)	
(i) Medium castings	1.5
(ii) Large castings	5

Table 5.2 *Machining allowances for various metals.*

5.1.2.3 *Draft Allowance*

After packing the sand around the pattern in moulding flasks, the pattern has to be withdrawn without damaging the mould cavity. The withdrawal of the pattern will be made easy by providing draft or taper allowance on the pattern. Draft is nothing but providing a slight taper on the vertical surfaces of a pattern. The amount of taper depends upon the shape and size of the pattern, moulding method and moulding material. Draft allowance is provided generally on internal as well as external surfaces. For external surfaces it is about 10 to 25 mm per metre whereas for internal surfaces it is about 40 to 65 mm per metre.

5.1.2.4 *Rapping Allowance*

Before withdrawing the pattern, the pattern is rapped or shaked for easy withdrawal. By doing so, the mould cavity gets slightly larger in size and causes the casting size also to increase. Hence to compensate this increase, the pattern is made initially slightly smaller than the required size. In case of small and medium sized castings this allowance may be insignificant whereas for large sized castings or where high precision is required, rapping or shake allowance is provided based on experience.

5.1.2.5 *Distortion Allowance*

A casting will usually distort or camber if it has irregular shape or if all parts of the casting do not shrink uniformly. For example a casting in the form of 'U' shape will shrink at the centre on cooling, whereas the legs will remain in fixed position. In this case to avoid distortion due to thermal stresses, the legs of 'U' pattern must converge slightly so that the legs will straighten and become parallel after the casting is cooled.

5.1.3 Pattern Design Considerations

The following factors should be taken into consideration while designing a pattern :

(i)   According to the size, shape and design of the casting, suitable allowances such as shrinkage, machining, draft, rapping and distortion should be provided wherever necessary on the pattern.

(ii)   Taking into account all the factors, proper material should be selected for making the pattern.

(iii)   All sharp edges and corners should be rounded on the pattern by providing a fillet. A concave connecting surface or the rounding of a corner caused by two intersecting surfaces or planes is called a fillet. These fillets or rounded corners help in easy withdrawal of pattern from the moulding box, smooth flow of molten metal into the mould cavity, minimizing internal stresses and strains and in producing sound castings.

(iv)   Changes in section thickness, i.e., whenever there is a change from light section to a heavy section or vice versa, should be gradual and uniform. Otherwise, internal stresses and strains will be induced in the casting and the casting becomes liable to crack.

(v)   The pattern should possess good surface finish since the surface finish of the casting much depends on the surface finish of the pattern.

(vi)   The pattern should be painted with some good quality paint if it has to be preserved for a longer time.

5.1.4 Types of Patterns

Following are the types of patterns commonly used in foundry depending upon the complexity of the job, the number of castings to be produced and the mould method.

(i)   single piece or solid pattern

(ii)   split pattern or two piece pattern

(iii)   multi piece or loose piece pattern

    (iv)   match plate pattern

    (v)   cope and drag pattern

    (vi)   sweep pattern

    (vii)   gated pattern

    (viii)   skeleton pattern

5.1.4.1 *Single Piece Pattern*

Pattern made in one piece without joints, partings or loose pieces is known as single piece or solid pattern. This type of pattern is inexpensive and simplest. A typical single piece pattern is shown in Fig. 5.1

    Single piece pattern is generally used for large casting of simple shape.

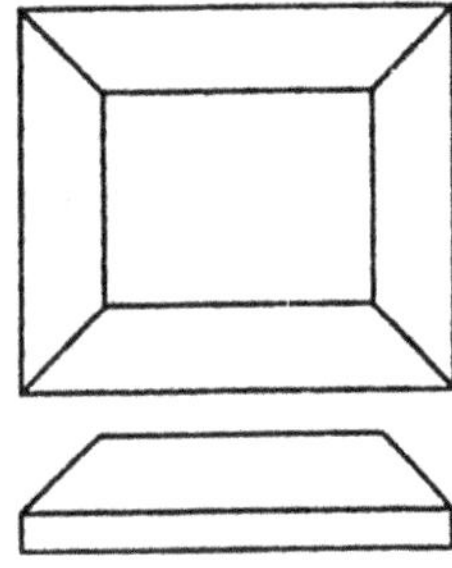

Fig. 5.1 *Single piece pattern.*

5.1.4.2 *Two Piece or Split Pattern*

Split patterns are used for intricate castings when use of single piece patterns are restricted due to difficulty in withdrawing the pattern. Split patterns are made in two pieces which are joined at the parting line by means of dowel pins. One part of the split pattern produce the lower half of the mould cavity and the other the upper half. A typical split pattern is shown in Fig. 5.2.

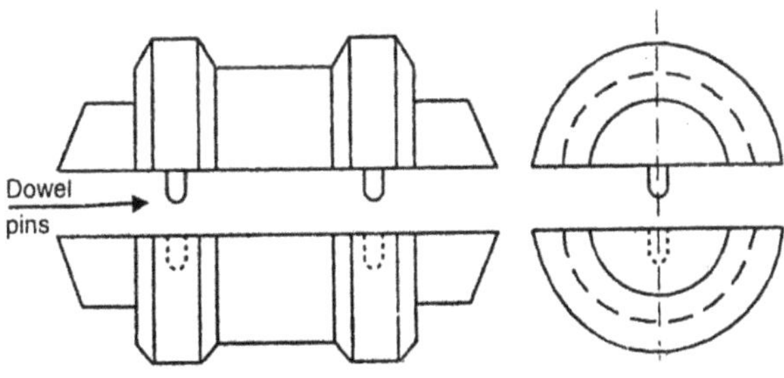

Fig. 5.2 *Split pattern.*

5.1.4.3 *Multi Piece or Loose Piece Pattern*

Multi piece or loose piece pattern is required when a one piece and a split pattern are unsuitable for withdrawal from the mould. In this case first the main pattern is usually removed first and then the loose pieces which are held by a wire are removed through the gap generated by the main pattern. A typical multi piece pattern is shown in Fig. 5.3 where A and B are loose pieces.

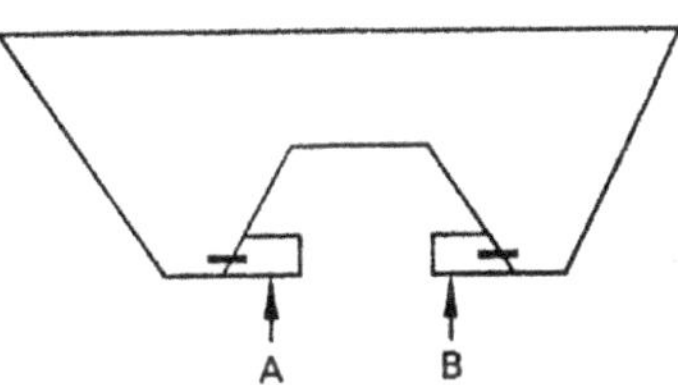

Fig. 5.3 *Loose piece pattern.*

5.1.4.4 *Match Plate Pattern*

Mass production of small castings utilizes match plate patterns. The cope and drag portions of number of patterns are mounted on opposite sides of plate conforming to the parting line.

Also the gating systems are almost always attached to the plate. The increased production rate possible with these patterns serves to compensate the increased cost. A typical match plate pattern is shown in Fig. 5.4

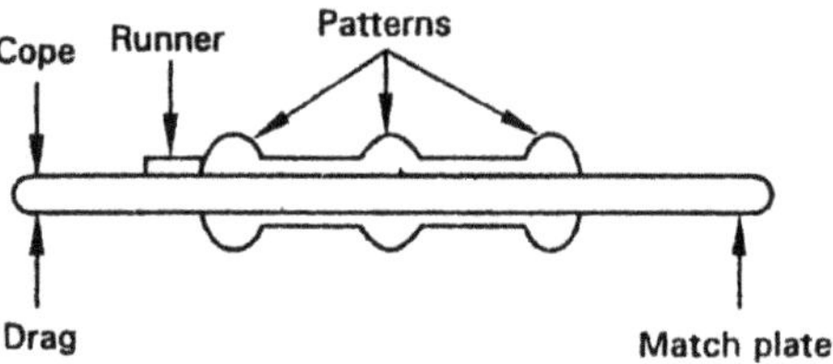

Fig. 5.4 *Match plate pattern.*

5.1.4.5 *Cope and Drag pattern*

These are similar to split patterns. The pattern is split on a convenient joint line and then the cope and drag halves of the pattern along with the gating systems are attached separately to the metal or wooden plates. Then the cope and drag halves are moulded separately by two moulders in a separate moulding box. The two moulds are finally assembled to form the complete mould. Cope and drag patterns are used for producing large size castings which as a whole can not be conveniently handled by one moulder alone.

5.1.4.6 *Sweep Pattern*

Sweep patterns are used for shaping moulds of symmetrical and regular shape particularly in large sizes. The sweep pattern consists of a wooden board having a shape corresponding to the shape of the required casting. As shown in Fig. 5.5, it is arranged to rotate about central axis. The sand is rammed in place and the sweep board is moved around its axis of rotation so as to give the moulding sand the required shape of the casting. Since no pattern is needed, this method is very economical.

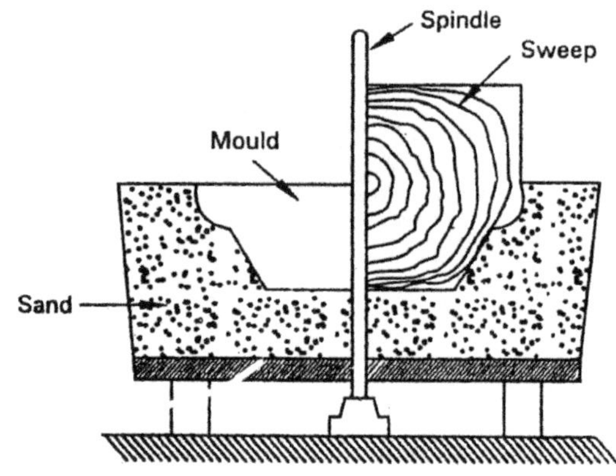

Fig 5.5 *Sweep pattern.*

5.1.4.7 *Gated Pattern*

Gated pattern is the one in which the gating system constitutes a part of the pattern. In this type, hand cutting of gates, runners etc., is not necessary or eliminated completely. As shown in Fig. 5.6 a set of patterns is arranged about a common runner and connected to it through ingates.

Gated patterns are usually made of wood or metal and are used for mass production of small castings.

5.1.4.8 *Skeleton Pattern*

When only a large number of small castings is to be made, it is not economical to make solid pattern since a large amount of timber is required. A skeleton pattern may, be used in such cases. Skeleton pattern is a wooden frame and rib construction which forms an outline of the casting. This frame work is filled with sand and the exact overall shape is obtained by firmly pressing the sand and removing the excess sand by a strike off board called as 'strickle board'. Skeleton patterns for round shapes such as pipes and cylinders are first prepared in halves whIch are then joined by dowels or by gluing. A typical skeleton is shown in Fig. 5.7

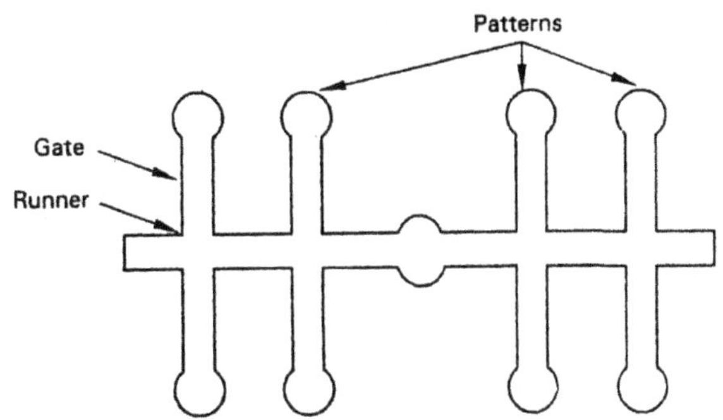

Fig 5.6 *Gated pattern.*

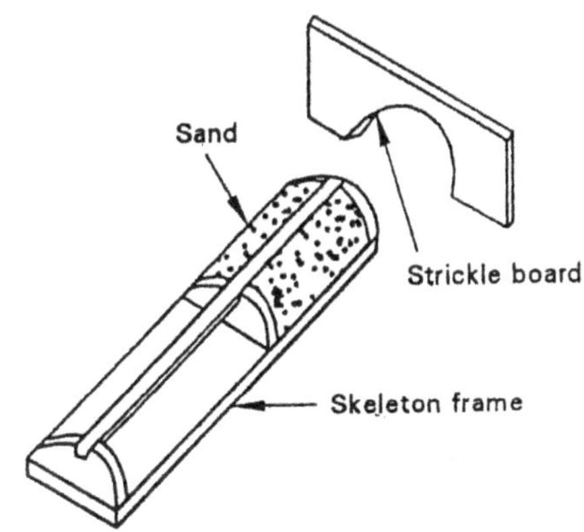

Fig 5.7 *Skeleton pattern*

5.1.5 Pattern Colour Code

To make the mould maker to identify clearly the functions, the patterns are usually painted with contrasting colours. The colour code is as follows :

  (i)   Red or orange on surfaces unfinished and left as cast.

  (ii)   Yellow on surfaces to be machined.

  (iii)   Black on core prints for unmachined openings.

  (iv)   Yellow stripes on black on core prints for machined openings.

  (v)   Green on seats of and for loose pieces and loose core prints.

  (vi)   Diagonal black stripes with clear varnish on stop offs to strengthen the weak patterns by reinforcing or portions of a pattern that form a mould cavity which is filled with sand before pouring.

5.2 MOULD AND CORE MAKING

Moulding material is the material which is used for mould making. The moulding material should be such that the mould cavity retains its shape until the molten metal solidifies completely.

Castings are normally made in permanent as well as temporary moulds. Permanent moulds are made of ferrous metals and alloys such as Grey cast iron, steel etc., whereas temporary moulds are made of refractory sands, resins, wax, carbon, ceramics, plaster of paris, etc. Of all these refractory moulding sand is the most common mould material due to its inherent properties and also it is cheap and easily available.

5.2.1 Moulding Sand

The principal constituent of moulding sand are as follows :

    (i)   Sand

    (ii)   Binder

    (iii) Water

    (iv) Additives

Sand

The principal constituent of sand is silica (sio_2) along with some impurities. Chemical composition of a typical sand is as follows :

SiO_2	–	89.4%
Al_2O_3	–	4.07%
Fe_2O_3	–	2.04%
TiO_2	–	0.2%
CaO	–	0.74%
MgO	–	0.50%
Na_2O	–	0.30%
K_2O	–	1.13%
Loss on ignition	–	1.20%

The chemical composition of sand gives an idea about the impurities like alumina, iron oxide, lime, magnesia, alkalies, etc. More the impurities less will be the refractoriness of sand.

The grain surface, shape and size of sand affects many of the sand properties. Smooth surfaced sand grains are preferred for moulding purposes since smooth surfaced grains possess higher sinter point.

The shape of sand grains may be rounded, angular, sub angular and compound. The sand grains used in foundaries are usually of mixed origin.

The grain size of sands influence the properties of moulding sands. Usually sand with wide range of grain size is preferred than the sand with narrow grain size distribution.

Binder

Binder is used to impart bonding strength to the moulding sand so that the mould does not lose its shape after ramming. The usual binders used in moulding sand are clay, sodium silicate and port land cement. Of these clay is most widely used. The clay used for foundry purposes are (i) kaolinite (ii) ball clay, (iii) fire clay (iv) limonite, (v) fuller's earth, and (vi) bentonite. Bentonite is the most common clay which is used as binder in moulding sands since it has high fusion point (1336°c) and majority of particles are of colloidal size (less than a micron). Due to their fine grain size, the bentonite particles have a large surface area and hence able to spread readily as a thin adhesive film around the sand grains without filling up the spaces between grains.

The binder content in moulding sand varies from 3 to 15 per cent but in certain special cases it may be present upto 50 per cent (loam moulding).

In some mineral deposits, clay and sand occur in mixed proportions so that this sand can be mined and directly used after adding water for moulding. This type of sand is referred as natural moulding sand. But in case of synthetic moulding sand which is prepared in foundry, the binder and water are mixed to the sand.

Water

Water content is one of the most important parameter affecting mould characteristics. Water content in the moulding sand generally varies from 1.5 to 8 per cent. The bonding action of clay (binder) depends on water. Water activates the clay in the clay - sand mixture and imparts strength and plasticity to the mixture (moulding sand).

The total amount of water can be divided into two classes.

(a) Adsorbed water : This is the amount of water which is adsorbed by the clay. Only that amount of water which is rigidly held (adsorbed) by the clay is responsible for developing strength in the moulding sand.

(b) Free water : This is the amount of water added in excess of adsorbed water. This free water acts as a lubricant, imparts plasticity to the moulding sand and thereby improves mouldability. But free water lowers the strength of the moulding sand.

Optimum water content which should be added to moulding sand depends on type of clay and its amount. Usually the amount of water added to develop the optimum properties can be determined experimentally.

Additives

Besides the three main ingredients, many other materials (called 'additives') may also be added to the moulding sand to enhance certain properties. The most import additives are listed below :

(i) *Cereals or corn flour* **:** These are added in amounts upto 2.0 per cent which increases green and dry strength, minimizes sand expansion defects (since cereals are completely volatilized by heat in the mould, leaving space between the sand grains), and improves collapsibility.

(ii) *Sea coal or coal dust* **:** This is finely powdered bituminous coal. This is added to moulding sands for cast irons, mainly to improve surface finish and ease of cleaning the castings. About 2 to 8 per cent of sea coal or coal dust are added.

(iii) *Pitch and Asphalt* **:** Pitch is obtained by distillation of soft coals at about 315°C whereas asphalt is a by product of petroleum distillation. These are added to improve hot strength and casting finish on ferrous castings.

(iv) *Graphite* **:** Synthetic or natural graphite is added from 0.2 to 2.0 per cent in order to improve the mouldability of the sand and the surface finish of the castings.

(v) *Silica Flour* **:** Pulverized silica, finer than 200 mesh, called silica flour is added in amounts upto 35 per cent to improve hot strength of the sand and to resist metal penetration by increasing the density of the sand.

(vi) *Wood flour, cereal hulls, carbonized cellulose etc.* **:** These cellulose materials are added to moulding sands in amounts of 0.5 to 2.0 per cent in order to minimize sand expansion defects since these materials burn out at elevated temperature. They can also improve collapsibility and flowability of the sand.

(vii) *Perlite* **:** Perlite, an expanded aluminium silicate mineral, is added in amounts of 0.5 to 1.5 per cent, to achieve better thermal stability of the sand. This is also used as a riser insulator.

(viii) *Fuel oil* **:** This is added in amounts of 0.01 to 0.1 per cent to improve mouldability of sand.

(ix) *Iron oxide* **:** Fine iron oxide is added in small percentages to improve the hot strength of the moulding sand.

(x) *Molasses and Dextrin* **:** These are added to improve dry strength and edge hardness of moulds.

5.2.1.1 *Basic Requirements (properties) of Moulding Sand*

The basic requirements of moulding sand are as follows :

(a) Retractoriness

(b) Permeability

    (c) Flowability

    (d) Adhesiveness

    (e) Cohesiveness

    (f) Green strength

    (g) Dry strength

    (h) Chemical resistivity

    (i) Collapsibility

    (j) Coefficient of expansion

(a) Refractoriness : The ability of the moulding sand to withstand high temperature of the molten metal without breaking down or fusing is known as 'Refractoriness'. Moulding sands having poor refractoriness will slag on the surface of the mould and hence smooth casting surface can not be obtained.

Refractoriness depends mainly on the silica content and the higher the silica content of the sand, the higher will be the refractoriness.

(b) Permeability : Molten metal absorbs certain amount of gases during melting in the furnace and during solidification these gases are evolved in the mould. Also moist sand and additives produce steam and other gases as soon as molten metal is poured into the mould. All these gases and steam must escape through the mould wall rather than through the metal. Otherwise they produce defects like gas holes and pores in the casting. Hence, the moulding sand must be sufficiently porous or permeable to allow these gases to escape through the mould walls. This property is known as permeability or porosity.

Permeability is a function of grain size and shape, clay content and the extent of ramming. Usually coarse sands with rounded grains exhibit more permeability. Also clay addition in lesser amounts and soft ramming improves permeability.

(c) Flowability : Flowability is the ability of the sand to behave like a fluid so that when rammed, it will flow to all portions of a moulding flask and distribute the ramming pressure uniformly. Usually flowability increases with increase in clay and moisture content.

(d) Adhesiveness : It is the property of moulding sand owing to which the sand particles must be capable of adhering to another body. In other words, they should cling to the sides of the moulding boxes. It is due to this property that the sand mass does not fall out of the moulding box when the mould is moved.

(e) Cohesiveness : The ability of sand particles to stick together is known as cohesiveness or strength of the moulding sand. If cohesiveness is not sufficient, the mould may collapse during conveyance, turning over, closing or pouring.

(f) Green strength : The cohesiveness or the strength of moulding sand in its moist or green state is known as green strength. A mould should possess adequate green strength, so that it will not distort or collapse during and after removing the pattern from the moulding box. Also dimensional stability and high accuracy cannot be achieved without adequate green strength.

(g) Dry strength : The cohesiveness or the strength of moulding sand in its dried or baked state is known as dry strength. A mould should possess adequate dry strength so that it can avoid erosion of mould wall during the flow of molten metal and also prevents the enlargement of mould cavity due to the metallostatic pressure of the molten metal.

(h) Chemical resistivity : The moulding sand should not react chemically with the molten metal, otherwise the casting shape will be distorted and smooth surface will not be obtained.

(i) Collapsibility : The sand mould must be collapsible, after the solidification of molten metal in the mould. This property of moulding sand allows free contraction of solidifying metal and thereby avoids tearing or cracking of the solidifying metal.

(j) Coefficient of expansion : Moulding sand should possess low co-efficient of expansion

5.2.1.2 *Testing of Moulding Sand*

Generally the following tests are performed on moulding sand :

    (a) Moisture content test

    (b) Clay content test

    (c) Grain fineness test

    (d) Refractoriness test

    (e) Strength tests

    (f) Permeability test

    (g) Shatter index test

    (h) Mould hardness test

(a) Moisture content test

Moisture content of moulding sand can be determined in any of the following ways:

    (i)   Moisture content can be determined by heating (105 to 110°C) a weighed moulding sand sample in an oven so that all the moisture evaporates. And then after cooling to room temperature in a desicator, it is reweighed. The loss in weight gives the moisture content which can be expressed as a percentag

    (ii)   Since the above conventional method is time consuming and tedious, direct reading instruments are often used to determine the moisture content quickly. Moisture teller is one such instrument, which makes use of the following reaction :

$$CaC_2 + 2\,H_2O \rightarrow Ca\,(OH)_2 + C_2H_2$$

The pressure of the acetylaene gas generated by the above reaction of calcium carbide with moisture of the moulding sand is indicated on a scale calibrated to read directly in percentage of moisture.

(b) Clay content test

Clay content is determined by A.F.S. method. According to A.F.S., clay in moulding sand is defined as particles (less than 20 microns) which when suspended in water fail to settle at the rate of one inch per minute. The steps involved in the clay content test are as follows :

1. Dry thoroughly a small quantity of moulding sand.

2. Take a sample of 50 grams of dry moulding sand and transfer the sample to the A.F.S. clay jar.

3. Add to it 475 C.C. of distilled water at room temperature and 20 c.c. of 2.5 to 3 per cent solution of sodium hydroxide.

4. By means of a stirrer which is driven by a motor, stirr the contents for 6 minutes.

5. Wash the sand which is on the stirrer into the jar and then fill the jar with distilled water to a height of 6 inches.

6. Allow the contents to settle for 10 minutes.

7. Then siphon off to a depth of exactly 5 inch, leaving a minimum depth of one inch of water in the bottom of the jar.

8. Add distilled water to the jar and the height is made upto 6 inches and then stirred.

9. Allow it to settle for 10 minutes and then siphon off 5 inches of water.

10. Add distilled water filling the jar to 6 inch height and then stirred.

11. Allow the contents to settle for exactly 5 minutes and then siphon off 5 inches of water

12. Repeat the steps (10) and (11) until the water is clear to a depth of 5 inches at the end of the 5 minutes period

By this method, the clayey matter which fails to settle at the rate of one inch per minute is removed from the moulding sand.

Now, the remaining sand in the jar is filtered carefully, dried for half an hour and weighed. The difference between this weight and the original weight (50 grams) gives the weight of the clay.

(c) Grain fineness test

The grain size is determined by grain fineness number. In this test, a dried 50 grams moulding sand sample is taken and clay content 5.6 grams i.e. 11.2 per cent is removed. The remaining 44.4 grams (88.8%) moulding sand sample is placed on top of a series of 11 sieves (sieve number and opening area given in Table 5.3) and shaken for 15 minutes. Then the amount of sand grains retained on each sieve is collected and weighed and expressed as the percentage of the total sand grade. Now these percentages obtained for each sieve are multiplied by the factor (multiplier) as shown in Table 5.3. The resulting products are added and divided by total percentage of sand grains retained. The quotient so obtained is known as the A.F.S. grain fineness number.

U.S. sieve equivalent No.	Openings mm	Amount retained on sieve		multiplier	product
		gram	%		
6	3.327	0.00	0.00	3	0
12	1.651	0.00	0.00	5	0
20	0.833	0.00	0.00	10	0
30	0.589	0.00	0.00	20	0
40	0.414	1.50	3.00	30	90
50	0.295	0.30	0.60	40	24
70	0.208	2.50	5.00	50	250
100	0.147	0.50	1.00	70	70
140	0.104	8.20	16.40	100	1640
200	0.074	13.00	26.00	140	3640
270	0.053	12.80	25.60	200	5120
Pan		5.60	11.20	300	3360
Total		44.40	88.80		14194

Table 5.3 Calculation of A.F.S. grain fineness number

$$\text{A.F.S. grain fineness number} = \frac{14194}{88.8}$$

$$= 160$$

(d) Refractoriness test

In this test a cylindrical sand specimen is fired at 1550°C for 2 hours. Then observe the appearance and dimensional changes. A good refractory sand retains specimen shape and shows very little expansion (7.0 percent or less). A poor refractory sand specimen will shrink and distort.

(e) Strength tests

The strength of moulding sand is determined by using universal sand strength testing machine. Usually moulding sand is tested for its compressive strength, tensile strength, shear strength and transverse strength.

For all these tests, the specimens are prepared under standard conditions using a laboratory sand rammer and its accessories. To prepare test specimens, a weighed amount of sand is filled in the specimen tube, and a rammer (6.35 to 7.25 kg weight) is allowed to fall on the sand three times from a height of 50.8 $\pm$ 0.125 mm. After the three blows, the specimen thus produced should have a height of 50.8 $\pm$ 0.8 mm. Usually sand of 145 to 175 grams is required to produce the specimen of this size. Using the attachments of sand rammer, different specimens with different size and shape are produced for various tests.

The different specimens (as shown in fig. 5.8) thus prepared are placed in universal sand strength testing machine and load is applied until failure occurs. The compression, tensile, shear and transverse strengths are measured directly on the scale provided on universal sand strength testing machine.

(f) Permeability test

Permeability is expressed in terms of the permeability number, which is defined as the volume of air in c.c. that will pass per minute through a sand sample of 1 sq. cm. in cross section and 1 cm height, at a pressure of 1 gram per cm^2. Hence,

$$\text{Permeability number} = \frac{vh}{pat}$$

where

v is the volume of air in c. c.

h is the height of the sample in cm

p is the pressure of air in gm/cm^2

a is the cross sectional area of the sample in cm^2

and t is the time in minutes.

As per the A.F.S. standard permeability meter 2000 c.c. of air is passed through a standard sand specimen (5.08 cm in height and 20.268 sq.cm. in cross sectional area) at a pressure of 10 gms per cm^2 and the total time is recorded in seconds. The permeability is calculated using the relation,

$$P = \frac{vh}{pat}$$

$$= \frac{2000 \times 5.08}{10 \times 20.268} \times \frac{1}{t\,(\text{minutes})}$$

$$= \frac{50.732}{t\,(\text{seconds})}$$

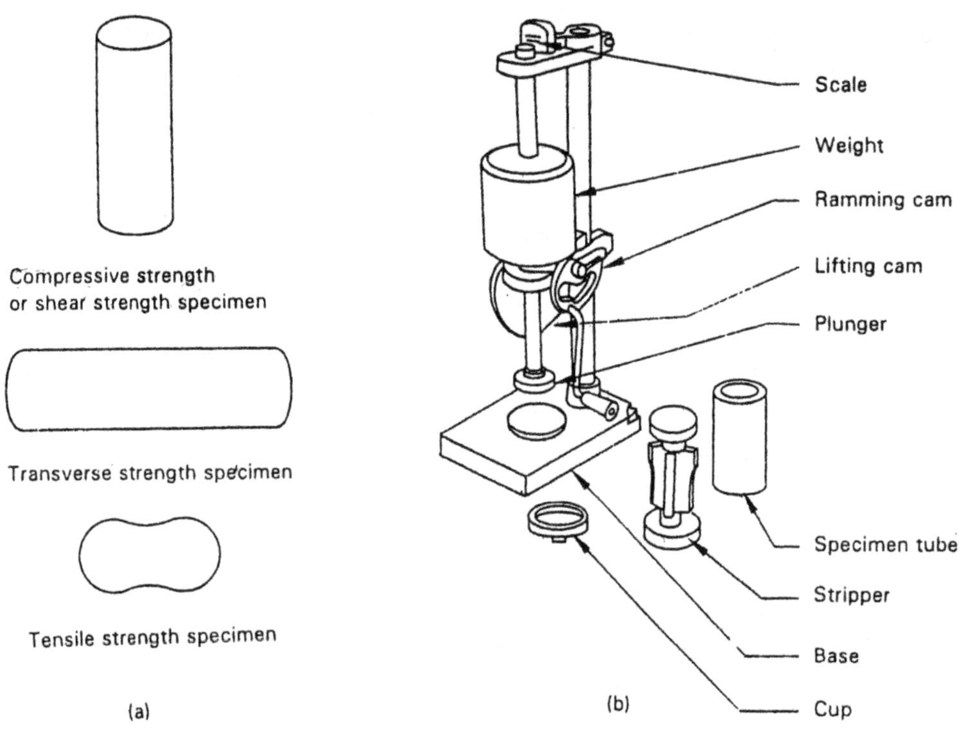

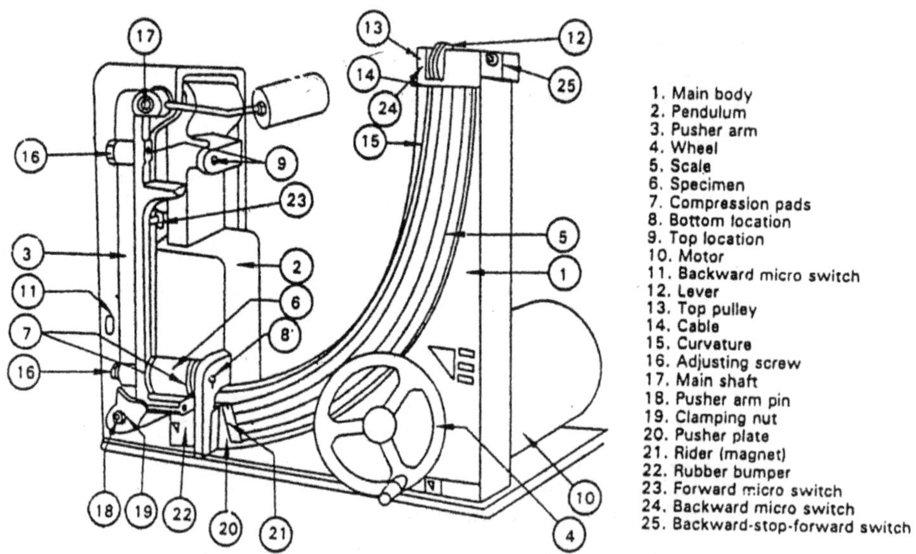

Fig. 5.8 *(a) Test specimens (b) Sand rammer (c) Pendulum type universal strength machine*

(g) Shatter index test

Shatter index is a measure of sand toughness, particularly the capacity of sand to withstand rough handling and strain during pattern withdrawal.

In this test, the A.F.S. standard specimen is rammed by 10 blows and then it is allowed to fall on a half inch mesh sieve from a height of 6 feet. The portion of sand retained on the sieve is weighed and is expressed as percentage of the total weight of the specimen. This gives the shatter index. Both extra low and extra high values, of shatter index, are not desirable to the mould.

(h) Mould hardness test

Mould hardness tester is used to determine the hardness of the mould. In a A.F.S. standard mould hardness tester, a half inch diameter steel ball is loaded with a spring load of 980 grams This ball is made penetrate into the mould surface and the penetration is indicated on a dial which is calibrated to read the hardness directly. Normally a mould surface which offers no resistance to the steel ball would have zero hardness value where as a mould which prevents completely the steel ball from penetrating would have a hardness value 100.

Mould hardness gives an idea about the ramming density of the mould.

5.2.2 Moulding Processes

Moulding processes are broadly classified based on the method and material used for moulding.

5.2.2.1 *Classification Based on the Method Used*

    (a) Bench moulding

    (b) Floor moulding

    (c) Pit moulding

    (d) Machine moulding

(a) Bench moulding

Bench moulding is favoured for small castings which are light in weight. In bench moulding the whole moulding operation is carried out by the moulder standing near the bench of convenient height. A minimum of two flasks namely cope and drag are required for bench moulding

(b) Floor moulding

Floor moulding is favoured for medium and large castings for which bench moulding is not suitable. In this method the floor itself acts, as the drag and the cope portion may be rammed in a flask and inverted on the floor (drag)

(c) Pit moulding

For very large castings (weighing 1 to 100 tons) which cannot be made in flasks because the pattern being moulded is too large, moulding is done in pits. Moulding pits are concrete lined square or rectangular shape holes in the moulding floor. In this method, the pattern is lowered into the pit and moulding sand is tucked and rammed under the pattern and up the walls to the parting surface. The cope of the pit mould is made by ramming the sand in cope flask. These moulds are always dried.

(d) Machine moulding

Hand moulding is a slow and laborious process suitable only for producing small quantities. Whereas for mass production hand moulding is not economical. In such cases machine moulding is generally employed. Machine moulding not only saves much time and labour but also produces castings with more accuracy, uniform in size and shape than those produced by hand moulding. Also less skilled labour can be employed to do the machine job.

The main important operation performed by machines are ramming the moulding sand by jolting, squeezing, jolt and squeeze and slinging.

Jolting : Jolt machines perform the action of jolting for ramming the moulding sand.

In this method, first the flask is filled with the moulding sand and then the flask is placed on the machine table. During the operation the machine table raises the flask a short distance and then the flask is allowed to fall freely under gravity. Due to this sudden action the sand gets packed and rammed. This action of raising and dropping the flask is called 'jolting'. The density and the

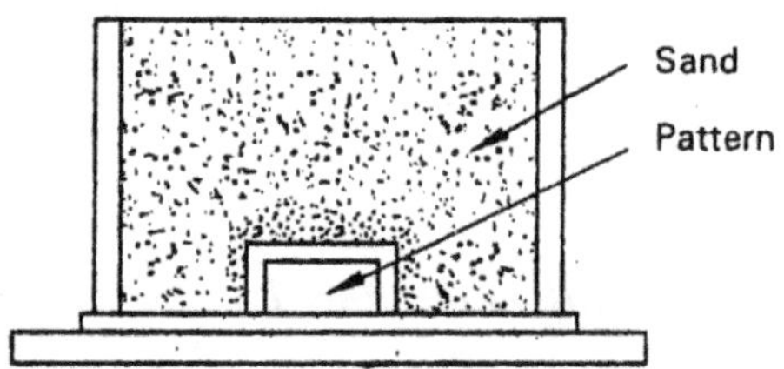

Fig. 5.9 Jolting Operation.

hardness of the mould depends on the height of the stroke, the amount of sand and the number of strokes.

The main draw back in jolting method is that the mould density is not uniform. The density is greatest around the pattern (as shown in Fig. 5.9) and minimum in the top layers. This necessitates hand ramming after the jolting action.

Squeezing : Squeezing machines perform the action of squeezing for ramming the moulding sand.

In this method, first the flask is filled with the moulding sand and then the flask is placed on the machine table. Then the moulding sand in the flask is squeezed between the machine table and the overhead squeeze board pneumatically or hydraulically with squeeze pressures which will vary from 3000 kg to 20,000 kg according to the size of the machine.

Like jolting in this method also the density and the hardness of the mould is not uniform. The density is greatest on the top of the mould from which the pressure is applied (as shown in Fig. 5.10) and the density decreases uniformly with the depth. Hence squeezing is preferred for moulds not more that 150 mm in depth.

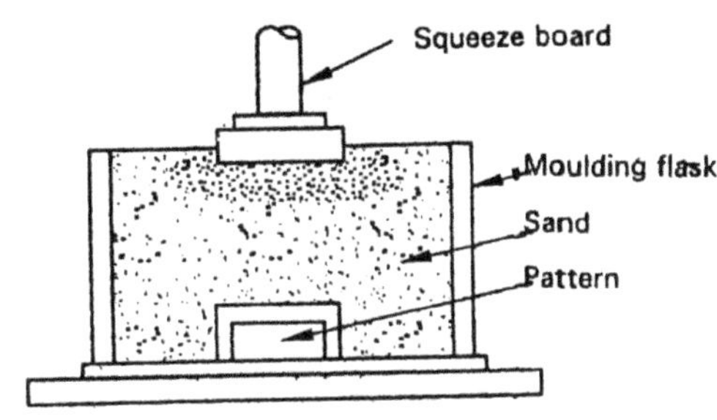

Fig. 5.10 Squeezing Operation.

Jolt and squeeze :

The draw backs of jolting and squeezing can be eliminated and uniform mould density and hardness can be obtained if these two operations are combined. The machines which perform the combined action of jolting and squeezing are referred to as jolt-squeeze moulding machines.

Slinging : Sand slinging machines, commonly known as sand slingers, fill and ram the mould simultaneously.

In these machines the ramming is achieved by the impact of sand which is thrown at a very high velocity with the help of a rotary impeller (Fig. 5.11). The impeller having cup shaped vanes rotates at high speeds. Moulding sand is fed by means of bucket elevators and conveyor belts to these cup shaped vanes. The sand is thrown at very high velocity into the mould, thus filling and ramming takes place simultaneously.

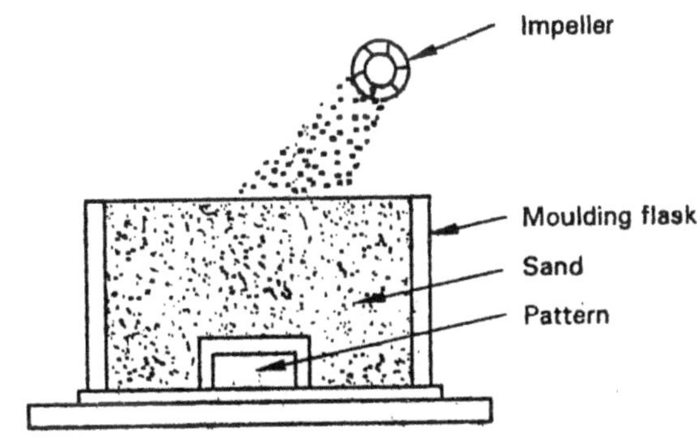

Fig. 5.11 Slinging Operation.

Slingers produce moulds with uniform density and this density can be controlled by varying the speed of the impeller. Slingers are faster in operation. The only draw back is that slingers are not suitable for small moulds and their initial cost is high.

5.2.2.2 Classification Based on the Material Used

    (i) Sand Moulding

        (a) Green sand mould

(b) Dry sand mould

(c) Skin dried mould

(d) Core sand mould

(e) Cement bonded sand mould

(f) Loam mould

(g) Carbon dioxide mould

(ii) Plaster moulding

(iii) Metallic moulding or permanent moulding.

Green Sand Mould

The sand is called 'green' because at the time of metal pouring the mould is in moist state i.e., it contains moisture. The main constituents of green sand mould are silica sand, clay, moisture and suitable additives.

Due to the presence of moisture, green sand mould lacks permeability, strengh and stability. Inspite of this draw back, green sand moulding is more popular and accounts for more than 90 percent of sand moulded castings because green sand moulding is most economical method.

Due to lack of permeability this method produces certain defects such as blow holes, gas holes etc., in castings. The surface finish obtained in this method is also not very smooth.

Dry Sand Mould

The sand is called 'dry' because at the time of metal pouring, the mould is in dry state. These moulds after preparation are baked in ovens to eliminate moisture.

The mould is prepared in the same way as that of green sand and then the mould surface is sprayed with molasses water which upon drying imparts greater hardness or refractoriness to the mould. The complete mould is then dried in an oven at 200 -300°C until all the moisture is eliminated.

Dry sand moulds considerably reduces the defects encountered in green sand moulds. But dry sand moulds are more expensive than green sand moulds due to baking operation. Dry sand moulds are often used when the mould require greater strength and hardness.

Skin Dried Mould

The expensive dry sand mould can be partially replaced with skin dried mould. In this case the full mould will not be dried but only the skin i.e., the surface layer of the mould to a depth of about 25 mm, is dried and hence the name skin dried mould. The skin may be dried by using gas torches or electrical heaters.

Skin dried mould gives one of the favourable properties of dry sand mould at very low cost of heating. However, skin dried mould is not as strong as dry sand mould and also skin dried mould must be poured shortly after drying the skin otherwise the moisture from the undried sand may penetrate the dried skin.

Core Sand Mould

In this case the complete mould is made by assembling cores. Core boxes are used instead patterns for making all parts of the mould.

Core sands usually consist of silica sand and organic binders which develop high strength after baking. The baking operation and organic binders which are expensive makes the process more costly. However the cost is justified in the intricate castings produced by core sand moulds. Core sand moulds are usually poured without a flask surrounding it. Core sand moulds possess high collapsibility and gives good surface finish to the castings.

Cement Bonded Sand Mould

Cement bonded sand mould mixture consists of silica sand, 10 per cent portland cement and 5% water. This requires less ramming since the setting action of cement gives much hardness and strength to the mould. Only after it sets a little, the pattern is withdrawn and then the mould is allowed to continue setting for about 72 hours before pouring molten metal. The moulds produce accurate and smooth surfaced castings.

The draw back with these moulds is that of poor collapsibility and hence it is very difficult to knock out of the flask.

Loam Mould

These moulds are built up with bricks and iron reinforcements and are given a thick coating of loam mortar all over and the shape of the mould is obtained with the help of sweep or skeleton pattern. Loam is a moist, plastic moulding sand containing about 50 per cent silica sand and 50 per cent clay. Before pouring the molten metal loam moulds are thoroughly dried.

Loam moulds are preferred only when one or two large castings are to be produced.

Carbon Dioxide Mould

The principle involved in CO_2 process is that if CO_2 gas is passed through a sand mixture containing sodium silicate as the binder. then hardening of sand occurs immediately as a result of the chemical reaction between sodium silicate and CO_2. The chemical reaction can be represented in simplified form as :

$$Na_2O.\, msio_2.\, xH_2O + CO_2 \rightarrow Na_2CO_3 + \underbrace{msio_2.\, xH_2O)}_{silica\ gel}$$

As given in the reaction the formation of silica gel which is the real hardening agent causes the hardening of sand. No drying or baking is required since the bonding strength obtained by the hardening action of silica gel is sufficient and molten metal can be poured immediately.

Preparation of CO_2 mould involves mixing of clean, dry silica sand (completely free from clay) and 3 to 7 per cent sodium silicate (binder) and some times coal dust, pitch, wood flour, and graphite are added to improve the collapsibility. This sand mixture is then treated with carbon dioxide gas at a pressure of about 1.4 kg/cm^2. Either metal or wood is used as a pattern material. Satisfactory results are obtained by spraying wood patterns with very thin layer of zinc followed by aluminium or brass.

Castings varying from few kgs in weight to more than 6 tons can be produced by this process.

Advantages :

    (i)   Shop cleanliness due to the absence of oven smoke

    (ii)   Considerable saving in labour since baking operation is not required and less skilled labour can be employed.

    (iii)   Better dimensional accuracy

    (iv)   The process can be mechanised and adopted even for mass production.

Limitations :

    (i)   Installation expenses for the supply of CO_2 gas is more

    (ii)   Poor collapsibiliy causes troubles

Plaster Moulding

In this process, a slurry consists of gypsum or plaster of paris ($CaSO_4$ .1/2 H_2O) 100 parts with additives like talc, asbestos fibres, silica flour etc. (to control contraction characteristics as well as setting time) and water 160 parts. This slurry is stirred to a creamy consistency and then poured over a metallic pattern confined in a flask. The mould is then vibrated so as to bubble out any air entrapped in the slurry and to ensure thal the mould is completely filled. The slurry is allowed to preset. Pattern is withdrawn and the preset mould is heated in an oven at about 200 °C so that the mould gets dehydrated. In the above manner cope and drag are made separately and are assembled for molten metal pouring.

Usually for this process a metallic or some other moisture resistant material is employed for patterns. Wooden pattems are not used because the water in the plaster raises the grains on them and causes difficulty in withdrawing the pattern.

Plaster moulding is used only for non ferrous alloys since the plaster possess low refractoriness. This method is very popular for ornamental castings, statues, jewellary etc.

Advantages :

(i) Plaster moulds imparts good surface finish and dimensional accuracy to the castings.

(ii) Plaster has very low thermal conductivity and hence slow and uniform rate of cooling of the casting is achieved which helps in eliminating the chilling tendency, thereby enabling thin sections to be easily cast.

(iii) Metal shrinkage can be controlled accurately

Limitations:

(i) Plaster moulds possess low permeability

(ii) If not dried properly evolution of steam occurs during metal pouring

Metallic or Permanent Moulding

In this, instead of sand mould, the molten metal is poured in a metallic mould. Metallic moulds are usually made of cast iron or steel which are not destroyed after every casting and hence they are called permanent moulds because of their long life. Generally these moulds called dies, are made in two halves and to obtain mould cavity they are clamped together. If the metal is introduced into the cavity under gravity it is called gravity die casting and if the metal is introduced into the cavity under pressure it is called pressure die casting.

Metallic moulds can be used both for ferrous and non ferrous alloys but they are more suitable for non ferrous alloys such as aluminium, zinc and magnesium because these alloys have low melting points.

Advantages :

(i) Castings possess fine grain structure due to rapid rate of cooling.

(ii) Good surface finish and dimensional accuracy is obtained.

(iii) Less skill is required

(iv) High production rate can be achieved

Limitations :

(i) Suitable for low melting point alloys due to low refractoriness of the mould as compared to sand mould.

(ii) Due to chilling effect, the surface of the casting becomes hard.

5.2.2.3 *Special Moulding Processes*

Various special moulding processes of commercial importance are developed and a few of them are as follows :

Shell Moulding

Shell moulding process was developed during second world war by a Scientist J. Corning for producing castings on mass production with smooth surface finish. This process was also called as Corning process or simple 'C' process.

The steps involved in shell moulding process are as follows :

(i)   A pattern (Fig. 5.12a) having the same profile as that of the casting is made with cast iron. It is then heated to 150 – 250 °C in an oven and is removed and sprayed on it a stripping agent usually 5 to 10 per cent silicone solution in paraffin or water to facilitate subsequent withdrawal of the shell from the pattern.

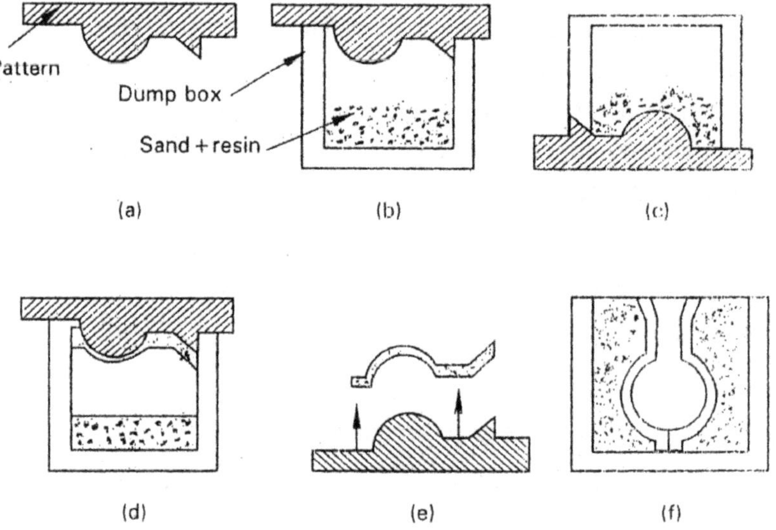

Fig. 5.12 *Shell Moulding Process.*

(ii)   The hot pattern is then clamped to the dump box (Fig. 5.12 b) which is mounted on trunnions. The dump box contains the shell forming medium i.e., the dry sand resin mixture. Usually zircon sand of rounded shape and free from organic matter is used. Resins (3 to 10 per cent by weight) which may be used are of thermosetting type such as phenol formaldehyde, urea formaldehyde, alkydes and polyesters. Resins may be used either as a powder in dry mixtures or as a liquid and then dried on the sand grains.

(iii) The dump box is inverted (Fig. 5.12 c) so that the sand mixture falls on the hot pattern face. The heat of the pattern causes the resin to soften and flow. Keep in this position for about 30 seconds so that the resin soften and fuses to form a uniform shell of about 1/4 inches thick on the pattern face.

(iv) Then the dump box is inverted back to its original position (Fig. 5.12 d) so that excess sand mixture falls back to the bottom of the dump box and leaving behind a thin shell adhering to the hot pattern surface. The thin shell is cured for about two minutes at 400°C.

(v) When the shell is cured completely, it becomes rigid and then stripped from the pattern by spring loaded ejector pins which pass through the pattern plate (Fig. 5.12 e)

(vi) With the help of clamps, resin adhesives or other devices, two such shells are fixed together to form complete mould and placed in a flask with backing sand to receive molten metal (Fig. 5.12 f)

Advantages :

(i) Very good surface finish and dimensional accuracy is obtained. Hence machining and cleaning cost is negligible.

(ii) Thin sections upto 1/16 inches can be cast.

(iii) Permeability of the shells is high.

(iv) Less skilled labour can be employed.

Limitations :

(i) Higher pattern cost

(ii) Higher resin cost

(iii) This process is limited to small sizes.

Ceramic Moulding

In order to produce metal castings of highest precision and very fine surface finish, this process employs chemical and ceramic slurries for moulds.

In this process a refractory slurry is prepared by mixing specially developed ceramic aggregates (critical blends of ceramic refractories) and a liquid chemical binder (modified alkyl silicate). The slurry so prepared is poured over the pattern confined in a flask. Slurry fills up all the cavities and recesses by itself and no ramming or even vibration of the mould is required. After about 3 to 5 minutes when the slurry sets to a rubbery consistency the pattern is withdrawn. The ceramic mass is removed from the flask and treated with a catalyst (sodium carbonate) or hardener for complete chemical stabilization. Then the mould is heated to about 980°C in a furnace to expel the liquid binder completely and the heating time depends on the size and sectional thickness of the mould. Now the mould is ready to receive the molten metal.

Advantages :

    (i)   Suitable for all types of cast metals and also for highly reactive metals such as titanium and uranium.

    (ii)   Any ordinary pattern of wood or metal may be employed.

    (iii)   Do not require risers, venting or chilling since the cooling rate is very slow.

Centrifugal Casting

In this process, the solidification of molten metal takes place while the mould is revolving i.e., the metal solidifies under the influence of centrifugal force. The application of centrifugal force assists feeding during solidification and also promotes the separation of dissolved gases, slag and any other impurities and makes the casting more dense and sound. Cylindrical parts and pipes are generally produced by this process. Since no gates or risers are used, the casting yield is increased to nearly 100 per cent.

Three methods of centrifugal casting are generally employed :

    (i)   True centrifugal casting

    (ii)   Semi centrifugal casting

    (iii)   Centrifuging.

(i) True Centrifugal Casting : In this the axis of rotation of the mould coincides with the axis of the casting and the formation of a central hole through the casting occurs by the centrifugal force without the use of a central core. The axis of rotation of the mould may be horizontal, vertical or at some other convenient angle between 0 and 90°. The axis of rotation mainly depends upon the L/D ratio of the tube, where L is the length of the job and D is the bore of the job.

For horizontal rotation : $\dfrac{L}{D} > 4$

For vertical rotation : $\dfrac{L}{D} < 1$

For any other angle : $4 > \dfrac{L}{D} > 1$

It is usual practice to spin a sand mould about a horizontal axis at a speed which will produce a centrifugal force of about 75 G where G is the force of gravity and for metal moulds it is 60 G.

(ii) Semi Centrifugal Casting : In this case the mould is rotated about its vertical axis and a central core is used to form the inner surface. Mould and core shapes will produce the shape of the casting and not by centrifugal force. The centrifugal force however assists in proper feeding of mould cavities like rim sections and eliminates porosity in the castings. This process is usually preferred for castings symmetrical about a central axis such as gear blanks, wheels and discs. The rotating speeds in this case are lower than true centrifugal casting and hence the impurities are not rejected towards the centre as effectively as in case of true centrifugal casting.

(iii) Centrifuging : This process is used for non-symmetrical castings. The main difference between true centrifugal or semi centrifugal and centrifuging is that in the case of true centrifugal or semi centrifugal casting the axis of mould coincides with the axis of rotation whereas in the case of centrifuging the axis of rotation does not coincide with the axis of the mould but it coincides with the axis of the whole assembly. In centrifuging, several mould cavities are located around central feeder and the molten metal is introduced at the central feeder which feeds to all the mould cavities through radial gates. Centrifuging is carried only vertical direction.

Moulds for Centrifugal Casting : Usually sand moulds, metallic moulds and graphite moulds are employed for centrifugal casting depending on the specific conditions.

Slush Casting

In this method hollow castings are produced without the use of cores. In this process, the molten metal is poured into a metallic mould and allowed to solidify a thin skin upto the required thickness and then the mould is inverted and the remaining molten metal is drained off. This way thin walled castings of desired thickness can be produced. Slush casting is mostly used for producing toys, statues and ornaments.

INVESTMENT CASTING

Investment casting process is also known as lost wax process or precision casting.

The term investment refers to a covering material. In this case a refractory mould covers the refractory coated wax pattern and in this process the pattern is expended.

The basic steps involved in this process are :

    (i)   Die making

    (ii)   Production of wax patterns

(iii)   Assembling of wax patterns

(iv)   Pre-coating of wax patterns assembly

(v)   Investment of coated wax patterns assembly

(vi)   Dewaxing and preheating

(vii)   Metal pouring

(viii)   Cleaning the castings.

The detailed discussion of these steps is as follows :

(i)   **Die Making :** In two ways the die, which is required for making the wax pattern, can be made.

   (a) Prepare a metal master pattern and then using this master pattern the die can be made by casting using low melting point alloys.

   (b) The Die can be made by machining the steel blocks.

Of these two ways, making the die by casting is easier than by machining.

(ii)   **Production of Wax Patterns :**

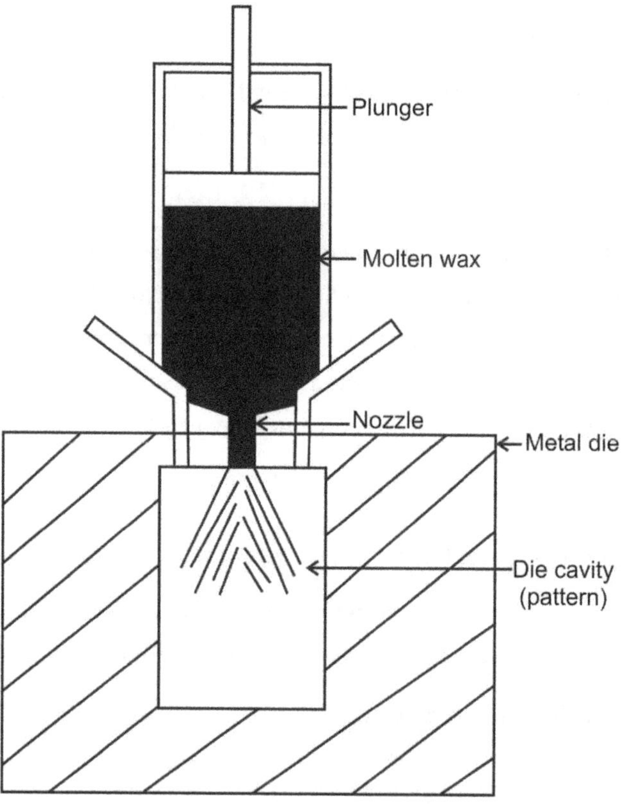

Fig. 5.13 Production of wax patterns.

Wax patterns and gating systems are produced by injecting the molten wax into the metallic dies.

Fig. 5.13 shows the production of wax patterns by injecting the molten wax (75 to 85 °C) at a pressure of 2 to 2.5 M pa into the die cavity. After injection the wax is allowed to solidify in the die cavity and then the wax pattern is removed from the die cavity.

(iii)   **Assembling of Wax Patterns :** With the help of hot spatula (broad rounded tool) the wax patterns are attached to wax gates and runners. This sort of assembly is called as 'tree'. The tree is shown in Fig. 5.14.

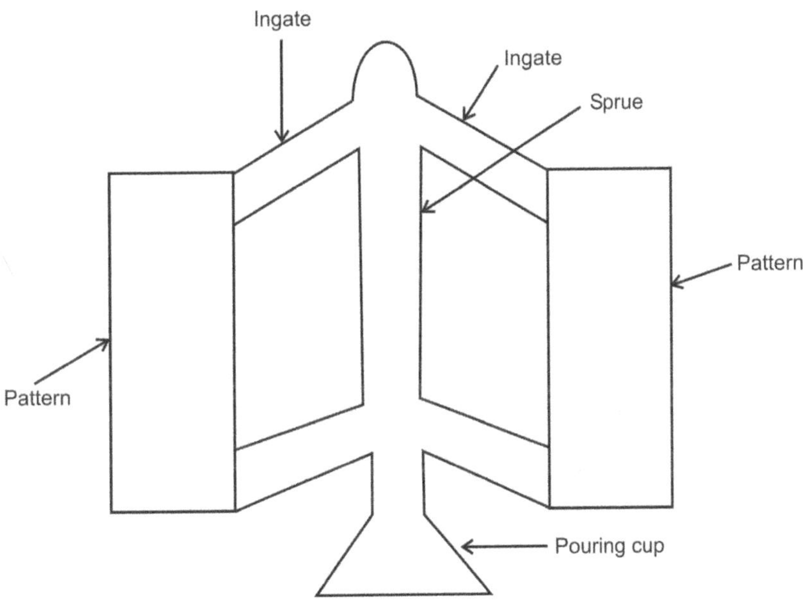

Fig. 5.14 Wax pattern assembly (Tree).

(iv)   **Precoating of Wax Pattern Assembly :** To give a refractory coating to the wax assembly, the wax assembly is dipped into a slurry which consists of 325 mesh silica flour suspended in ethyl silicate solution. The viscosity of the slurry should be made suitable so that the refractory coating will be uniform after drying.

(v)   **Investment of Coated Wax Pattern Assembly :** The coated wax pattern assembly is attached to a wax coated plate and then the whole assembly is surrounded by a steel flask. Now the slurry of investment moulding mixture is poured around the pattern. To ensure proper consolidation of the investment slurry around the patterns, the whole assembly is vibrated at a speed of 600 times per minute for few minutes. The mould is then allowed to set in the air.

Typical investment moulding slurries are :

    (a)   95 per cent sand, 27-30 per cent water, 5 per cent alumina cement.

    (b)   Sand + 3 percent or more ethyl silicate or sodium silicate.

(vi)   **Dewaxing and Preheating :** To obtain the mould cavity, wax is melted out of the hardened mould by heating to a temperature of 100 to 150 °C. The wax may be reclaimed and reused.

After wax melting, then the moulds are preheated to a temperature of 800 to 1000 °C to eliminate the wax completely and to get a wax free mould (Fig. 5.15).

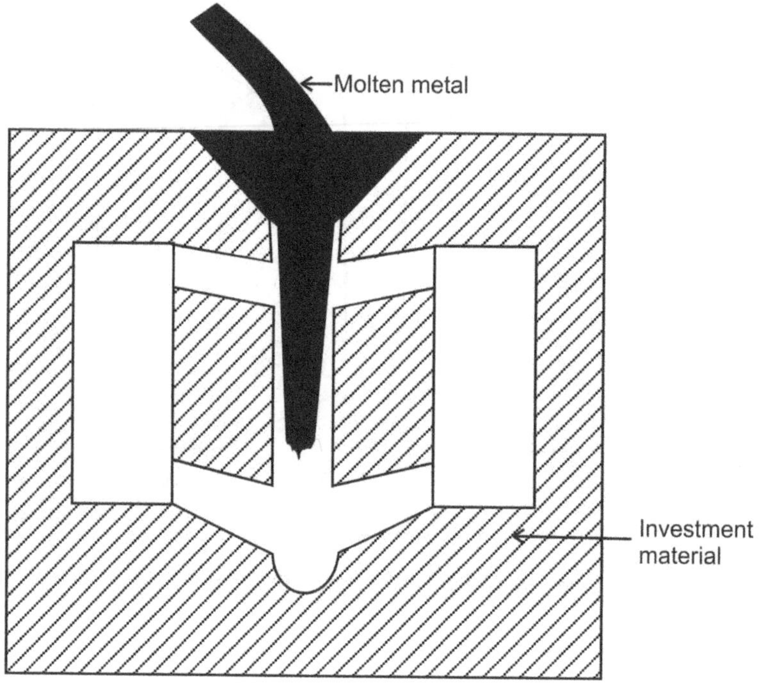

Fig. 5.15 Mould cavity formed after dewaxing.

(vii)   **Metal Pouring :** After preheating the mould, the molten metal is then poured into the hot mould to get the desired castings.

(viii)   **Cleaning the Castings :** After solidification of the molten metal, the castings are then separated from the mould and are cleaned and inspected before handing over to the customer.

Advantages of Investment Casting :

   (a) Better dimensional accuracy, better surface finish can be obtained because fine grain slurry is used for precoating the wax pattern assembly.

   (b) Complex shapes can be produced because here there is no with drawal of the pattern; pattern is removed by melting it.

   (c) No machining of castings is required and hence hard to machine alloys may be cast by this method.

   (d) Thin sections of the order of 0.75 mm can be cast.

Disadvantages of Investment Casting :

   (a) The process is expensive because more labour is required to prepare the pattern and mould.

   (b) Process is relatively slow.

   (c) It is suitable only for small size castings

   (d) Pattern is expendable

Applications :

   (a) Difficult to machine alloys such as turbine blades can be cast.

   (b) Parts for sewing machines, locks etc., can be produced.

   (c) For making jewellary and art castings

   (d) Wave guides for radars, bolts and triggers for fire arms, valve bodies, impellers for turbo chargers can be produced.

MERCAST PROCESS

In this process frozen mercury is used to prepare the pattern instead of wax.

Frozen mercury patterns are first prepared by pouring the mercury into the metallic moulds and allow it to freeze at a temperature of $-20°c$.

The mercury patterns so formed are removed from the mould and dipped into a ceramic slurry to form a uniform coating of about 3 mm.

The mercury is then melted and removed at room temperature. The shell so formed is then dried and heated to get harden.

The shell is then placed in a flask with backing sand and then the molten metal is poured to get the required castings.

Advantages :

   (a) Better dimensional accuracy and surface finish.

   (b) Both ferrous and non ferrous metals can be cast.

High cost is the disadvantage of this process.

ANTIOCH PROCESS

This process is an extension of plaster casting. Here an attempt is made to improve the permeability of the plaster mould.

In this process, a creamy slurry is prepared by adding water to a dry mixture of gypsum, sand, talc and small amounts of sodium silicate, portland cement. This slurry is then poured around the patterns which is confined in a separate cope and drag flasks. After the initial setting the patterns are withdrawn and then the cope and drag flasks are assembled in green condition and dried for about 6 hours in air. Then the moulds are placed in steam autoclave (at a pressure of 0.7 kg/cm^2). This autoclaving develops a more permeable structure in the mould. After autoclaving, the moulds are dried in air and final curing is done in an oven at a temperature of 230 °C for about 10 hours. This curing in oven further improves permeability by expelling the free and combined water.

Now the mould is ready to receive the molten metal.

Advantages :

    (a) Good surface finish and dimensional accuracy

    (b) Better mould permeability

COLD SET PROCESS

Two types of cold set process are in use.

These are :

    (i)   Cold set process using gaseous catalyst and

    (ii)   Cold set process using liquid catalyst.

The detailed discussion of these two processes is as follows :

(i) Cold Set Process Using Gaseous Catalyst

Organic resin binders are used in this process. The main advantage of this process is that heating of patterns like in shell moulding is not required. The curing of the mould takes place in cold condition by just passing a gaseous catalyst through the blown mass of sand in the core box.

In this process, fine dry sand is mixed with two binders – polyisocyanate and alkyl phenolic resin. This mixture is blown into a core box and then injecting tri ethylamine vapour (catalyst) through the blown mass of sand.

                            polyisocyanate

Drysand + two binders <   +   triethyl amine vapour

                        Alkylphenolic resin   (catalyst)

Since triethylamine vapour is toxic, the gas after passing through the core box, it is introduced into a scrubber solution where all of it is dissolved without entering into the atmosphere.

After treating with the catalyst, the hardened cores will be removed from core boxes.

Advantages :

(a) No heating of patterns is required.

(b) The process is excellent for mass production.

Disadvantages :

(a) Special expensive equipment is required for blowing the catalyst gas.

(b) The catalyst gas is poisionous.

(ii) Cold Set Process Using Liquid Catalyst

Cold curing processes using liquid catalyst have been developed to obtain better dimensional accuracy and better surface finish.

These processes are broadly of two types viz., alkyl isocyanate process and furan process.

(A) Alkyl Isocyanate Process : This process can be used both for making moulds and cores.

In this process, fine dry sand is mixed with two binders – alkyl resin and polyisocyanate and a liquid catalyst such as cobalt or MgO.

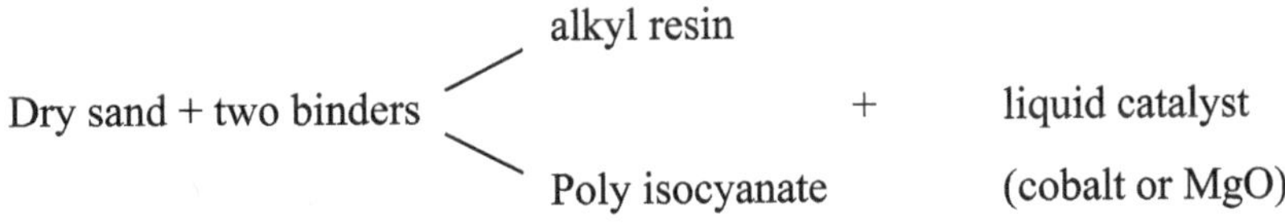

This mixture is filled in the moulding box or core box and then lightly compacted and left out in the air.

Then the hardening reactions will occur which causes the sand mixture to set quickly and forms a hard and strong mass.

The setting time varies from 20 - 40 minutes. After setting the moulds acquire good strength and permeability and are ready to receive the molten metal.

Advantages :

(a) Good surface finish and dimensional accuracy.

(b) Suitable for large castings.

(c) No generation of heat and hardening reactions take place in the cold state.

(B) Furan Process : This process is restricted for core making due to its high cost but it can also be used for mould making when good surface finish is desired.

The usual types of furan resins are :

(a) Urea furfuryl alcohol formaldehyde and

(b) Phenol furfuryl alcohol formaldehyde.

In this process, dry sand is mixed with either urea furfuryl alcohol formaldehyde or phenol furfuryl alcohol formaldehyde and a liquid catalyst, in this case phosphoric acid.

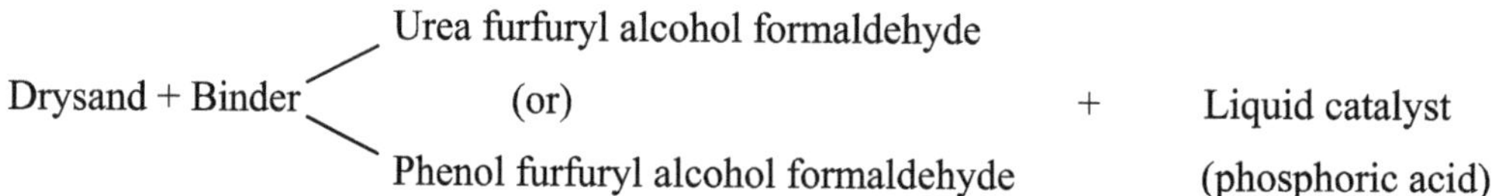

This mixture which is called as furan sand is placed in the mould box or core box and allowed it to set. Then immediately phosphoric acid causes hardening reaction to occur due to polymerization. This hardening reaction in the cold state makes the mould to get harden.

The urea type furan sand is cheaper and is generally suitable for gray cast iron.

For steel castings phenol type furan sand is suitable because the presence of nitrogen in urea type furan sand causes the formation of pin hole porosity in steel castings.

Advantages :

(a) Furan sands have high collapsibility and hence cleaning costs are much lower.

(b) No baking of moulds or cores is required.

(c) Ramming requirements are drastically reduced.

(d) Process is non toxic and non inflammable.

The disadvantage is that furan resins are costly which makes the furan process more expensive.

5.2.3 Cores

Cores are generally required for making cavities and hollow interiors of the castings which otherwise could not be obtained by the pattern alone or the mould proper. By using cores intricate castings can be easily produced.

5.2.3.1 *Core Sand*

The environment in which the core exists is much different from that of the mould. Generally cores are surrounded in all sides by the molten metal and hence they have to withstand the severe action of the molten metal. Therefore core sand should possess certain special characteristics as compared to moulding sand. These are :

(i)   As compared to moulding sand, the core sand should have high permeability so that the core gases can escape easily through the limited area of the core prints.

(ii)   Core sand should have high refractoriness since cores are subjected to the severe action of molten metal

(iii) Core sand should be collapsible in nature so that the cleaning of the castings becomes easy.

(iv) Core sand should not possess any such material which may produce gases while the cores come in contact with the molten metal.

The main constituents of the core sand are pure sand and a binder. Generally pure silica sand with coarse grain size is used for high refractoriness and for still higher refractoriness zircon or olivin is sometimes preferred. For core making organic binders such as core oil, molasses, cereal binder, dextrin are preferred. These binders produce small amount of gases when come in contact with molten metal.

5.2.3.2 Production of Cores

Basically production of cores consists of the following steps :

(i) core sand preparation

(ii) core making

(iii) core baking

(iv) core finishing

5.2.3.2.1 Core Sand Preparation

For uniform and better properties core sands are generally mixed with the help of muller type mixer or core sand mixer where uniform and homogeneous mixing occurs. Muller type of mixer is suitable for core sands containing cereal binder whereas core sand mixer is suitable for all types of core binders.

5.2.l.2.2 Core Making

Core making machines are classified as :

(i) Core blowing machine

(ii) Core shooting machine

(iii) Core ramming machine

(i) Core Blowing Machine

For rapid production of small and medium sized cores, the core blower is often employed. The basic principle involved in core blower is to blow the core sand into the core box at high velocity by using compressed air (at 5 to 7 kg/cm^2 pressure) (Fig. 5.16). Here the shapipg and ramming is done simultaneously in the core box since the sand comes with high kinetic energy.

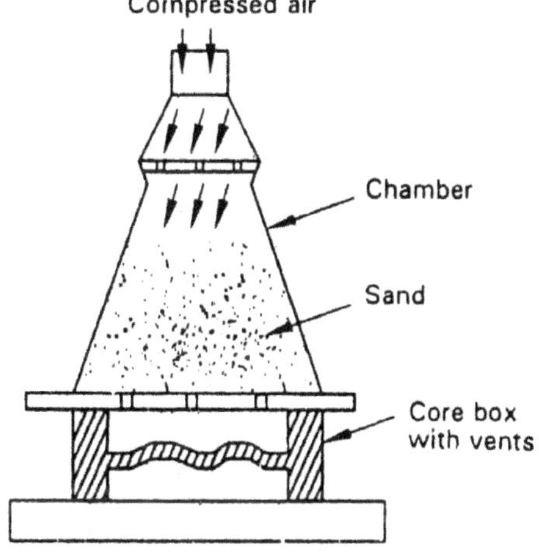

Fig. 5.16 Core Blower.

The air incoming with the sand is exhausted through vents provided in the core box. Core boxes which are used in core blowing machines are usually made with aluminium.

The usual problem in core blower is the channelling of sand in the chamber. This may be prevented to certain extent by agitating the sand in the chamber.

(ii) Core Shooting Machine

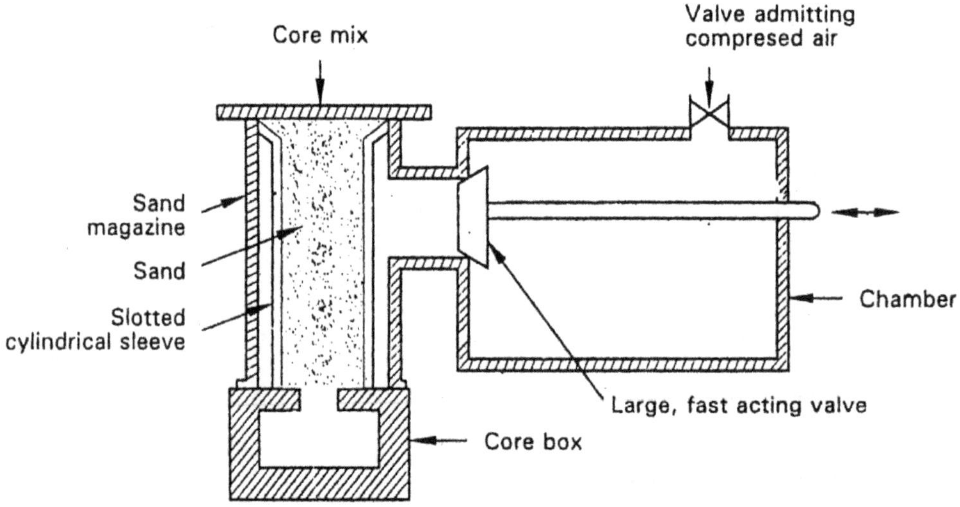

Fig. 5.17 *Core shooting machine.*

As shown in the Fig. 5.17, compressed air is admitted to a chamber and the entrance valve is then closed. Next a large fast acting valve opens and admits this predetermined amount of air into the sand magazine in the space outside of the slotted cylindrical sleeve. Pressure is applied to completely outside the core sand mix and thus moving core sand mix downward as a body into the core box. The core sand is essentially extruded into the core box.

In this method core box wear is negligible, less air is required and core box venting problems are greatly reduced.

(iii) Core Ramming Machine

Cores can also be produced by ramming core sands in the core boxes by machines based on the following principles.

    (a) jolting

    (b) squeezing

    (c) slinging

The principle of the three machines is discussed in section 5.2.2.1

5.2.3.2.3 Core Baking

After the cores are prepared, they are placed on supporting plates and transported to baking equipment. The main purpose of baking is to drive off the moisture and harden the binder. The core baking equipments are of two types :

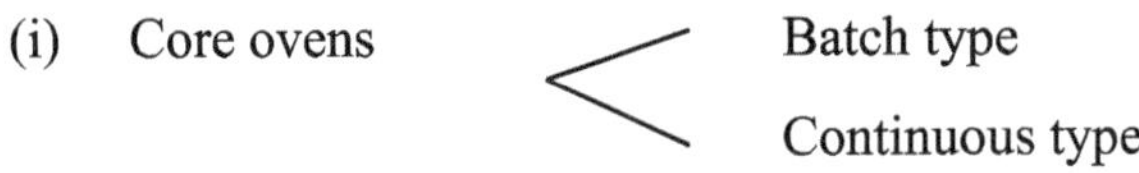

    (i)   Core ovens

    (ii) Dielectric bakers

5.2.3.2.4 Core Finishing

Certain finishing operations are carried after baking the cores and then they are set finally in the mould. First, by rubbing or filing the fins, bumps or other sand projections are removed from the surface of the cores. Next the dimensional inspection of cores is carried. Then the core surfaces may be coated (dressed) with a heat resistant paint to improve their refractoriness and surface finish. Core dressing may be applied either by dusting in case of solid powder or by dipping, spraying and brushing in case of liquid coatings.

5.2.3.3 Types of Cores

Cores are of different types depending upon their shapes and position in the mould. The various types of cores are as follows :

(i) Horizontal Core

This is the most common type of core and is mostly in cylindrical form. As shown in the Fig. 5.18, this core is placed in the mould horizontally. The two ends of the core rest in the seats provided by the core prints on the pattern. Horizontal core is placed at the parting line so that one half lies in the drag and the other half lies in the cope.

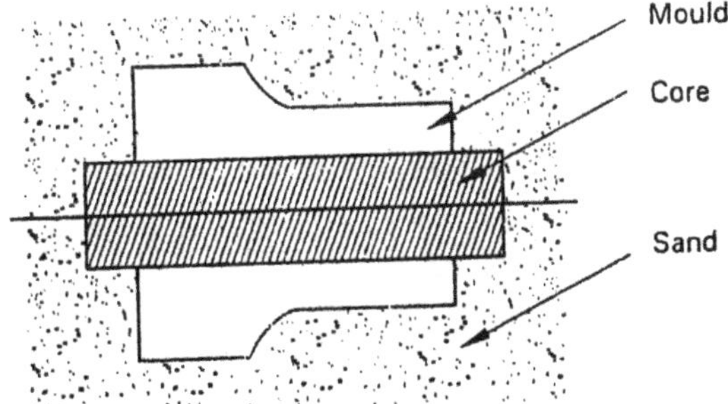

Fig. 5.18 Horizontal Core.

(ii) Vertical Core

The core placed in a vertical position in mould as shown in Fig. 5.19 is known as vertical core. Usually the upper end of the core is placed in the cope and the lower end in the drag.

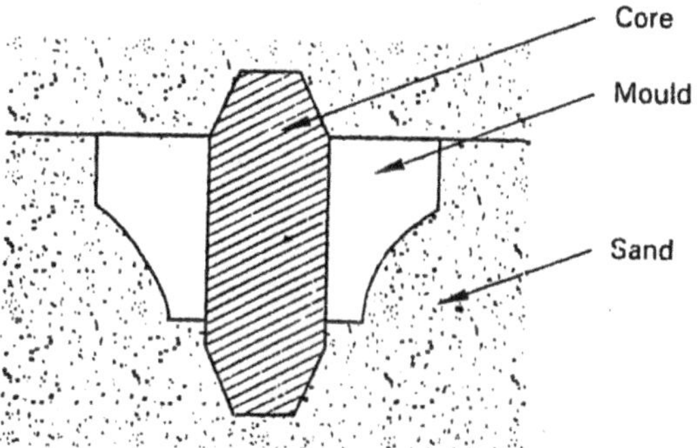

Fig. 5.19 Vertical Core.

(iii) Balanced Core

A balanced core is used when the casting has opening only on one side and only one core print is available (Fig. 5.20). In this case the core print should be sufficiently large in order to support the weight of the core and it should be able to withstand the force of buoyancy of the molten metal which surrounds it. In case the core is sufficiently long, it may often be supported at the other end with the help of a chaplet. These chaplets are made of the same metal of which casting is made and they finally go into the casting.

(iv) Hanging Core

If the core hangs from the cope and does not have any support at the bottom in the drag, it is referred to as hanging core (Fig. 5.21). In this case, it is often necessary to fasten the core with a wire or rod extending through the cope.

(v) Wing Core

Wing core is used when the axis of the desired hole does not coincide with the parting line of the mould. In such cases the core should be placed either above or below the parting line. Fig. 5.22 shows the wing core which is placed below the parting line.

(vi) Kiss Core

When core prints are not available for the core to rest, the core is held

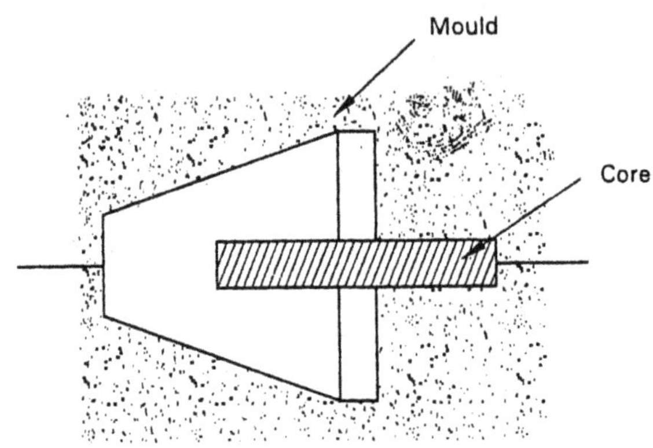

Fig. 5.20 Balanced Core.

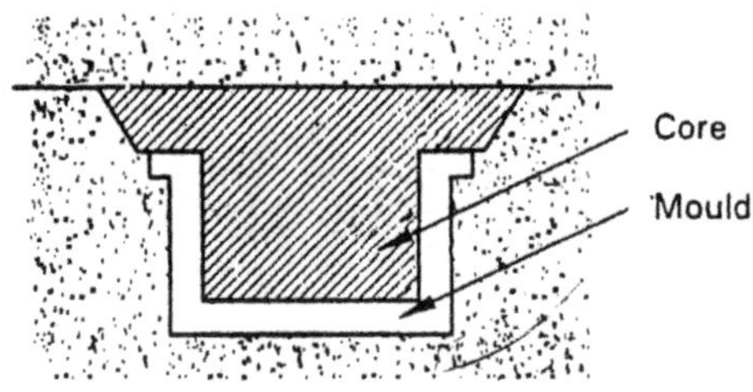

Fig. 5.21 Hanging Core.

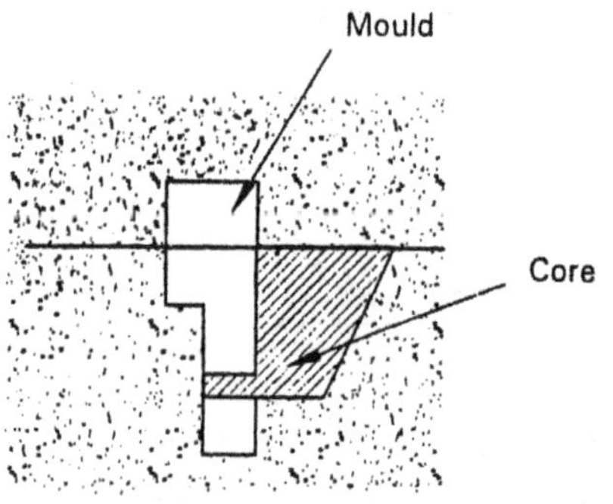

Fig. 5.22 Wing Core.

in position between the cope and drag simply due to the pressure of the cope (Fig. 5.23). This type of core is known as kiss core. Kiss cores are used when many holes are needed in the casting and the dimensional accuracy with respect to the location of the holes is not important.

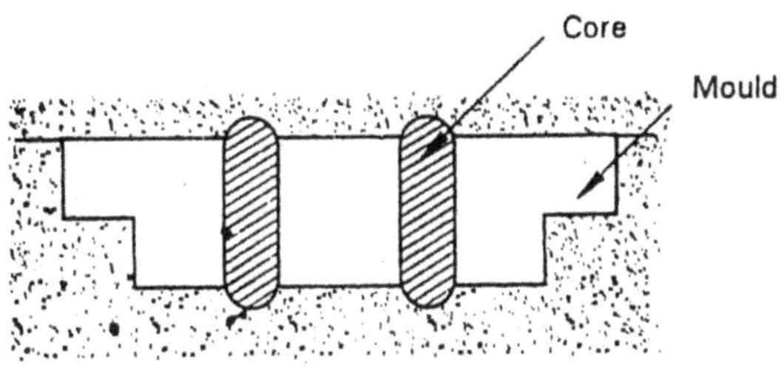

Fig. 5.23 *Kiss Core.*

5.2.3.4 *Core Prints*

In order to make the cores to rest securely and to place them in correct position in the mould cavity, core prints are provided. The core prints held the core rigidly so that is is not shifted during metal pouring and after that. Cores in the mould should be firmly secured so that they can withstand the buoyancy effect of the molten metal.

5.2.3.5 *Chaplets*

When core prints are inadequate, chaplets are provided to support the cores without shifting or sinking. Chaplets are metallic supports which are placed between mould and core surfaces. Chaplets should be of the same composition as that of casting and when molten metal is poured chaplets melt and go into the casting. Usually the chaplets before setting in the mould, should be thoroughly cleaned of any dirt, oil. rust or grease. Fig. 5.24 shows the use of chaplets.

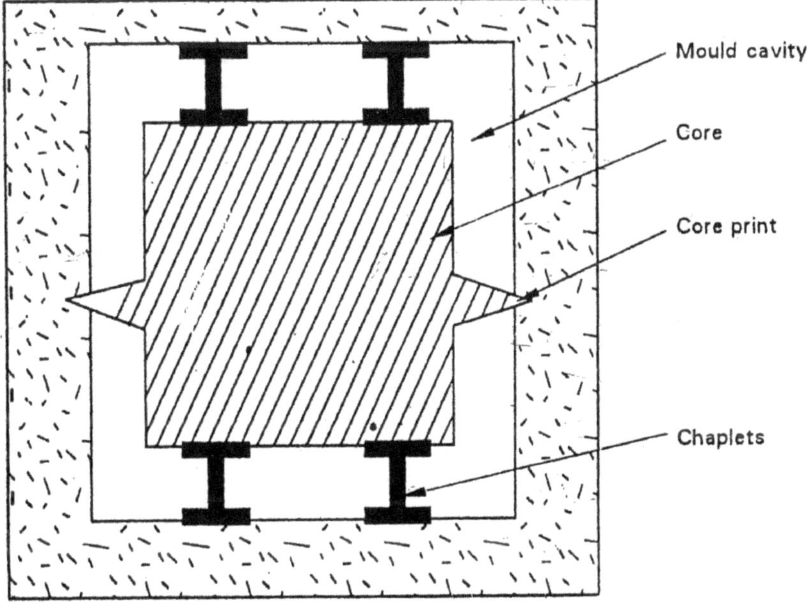

Fig. 5.24 *Core supported between chaplets.*

5.3 MELTING OF METALS

The metal to be cast should be in the molten condition before pouring into the mould. The metals obtained by the extraction processes, which are in molten condition, can not be poured directly into the moulds. This is because the metal so obtained is not in sufficiently refined state and also practically it is difficult to pour the huge quantities of molten metal, obtained by extraction processes, into the moulds. Therefore the metal obtained by extraction processes is first cast into some regular forms such as pigs and ingots, and these forms are remelted in foundries to pour into the moulds to obtain the required objects. For melting purpose, wide range of furnaces are employed in foundries. From the subject point of view the following furnaces are discussed below :

5.3.1 Cupola Furnace

The cupola furnace is the standard melting unit of the iron foundry. It converts cold iron scrap or pig iron into usable cast iron. Cupola gives continuous supply of molten metal and it is the cheapest melting unit.

5.3.1.1 *Description of a Cupola*

Figure 5.25 shows the cross sectional view of a cupola. The important parts of a cupola are as follows :

      (i)   **Shell :** The vertical, cylindrical shell is made of steel plates riveted or welded construction and is lined inside with either acid or basic refractory bricks. The shell is mounted either on a brick work foundation or on steel columns. At the bottom of the shell, a drop bottom door is provided through which debris consisting of coke, slag, etc., can be discharged at the end of a melt.

      (ii)   **Tuyeres :** Tuyeres are the openings through which air for combustion of fuel is forced into the cupola under high pressure.

      (iii)   **Charging Platform :** Charging platform is provided at about 0.3 metre below the bottom of the charging door.

      (iv)   **Charging Door :** Charging door is provided at about 3 to 6 metres above the tuyeres depending upon the size of the cupola. Through the charging door metal, coke and flux are fed into the furnace.

      (v)   **Spark Arrester :** The exhaust gases of the cupola carry a huge quantity of incandescent dust. A spark arrester is fixed at the top of the cupola to catch this dust and to prevent fires.

5.3.1.2 *Zones in a Cupola*

The various zones in a cupola which are shown in Fig. 5.25 are as follows :

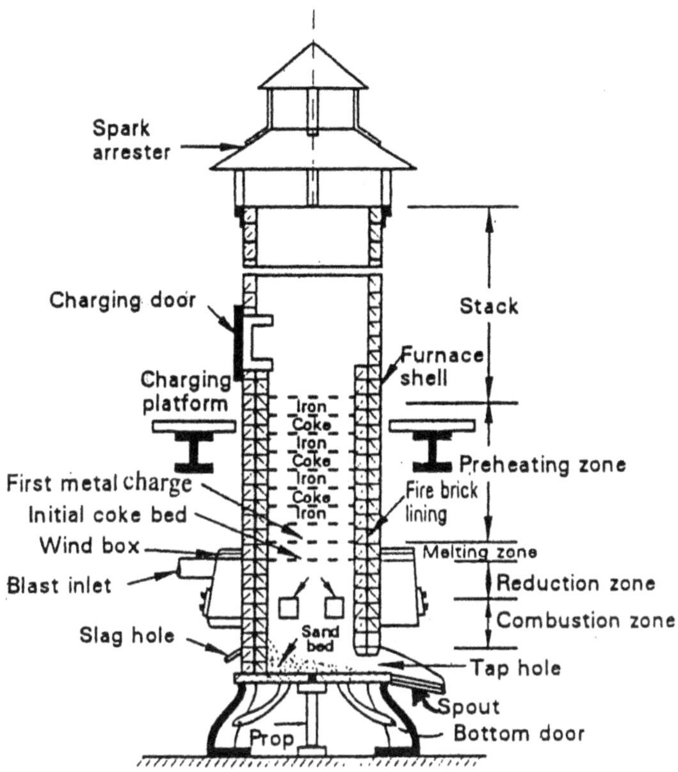

Fig. 5.25 Cross section of cupola furnace.

(i) **Hearth Zone :** This is the portion located between the bottom of the tuyeres and the bottom of the cupola. Slag and molten metal are accumulated in the hearth where slag hole and tap hole are provided to remove slag and molten metal from the furnace.

(ii) **Combustion Zone :** This zone is located above the hearth where intensive combustion of the fuel occur by the oxygen of the air blast and produces lot of heat in the cupola according to the following reaction

$$C + O_2 \rightarrow CO_2 + heat \text{ (exothermic)}$$

(iii) **Reduction Zone :** This zone is located above the combustion zone to the top of the coke bed. In this zone, the reduction of CO_2 to CO occurs which is an endothermic reaction thereby absorbs heat.

$$CO_2 + C \rightarrow 2CO \text{ (endothermic)}$$

(iv) **Melting Zone :** Melting zone is the first layer of iron above the coke bed. Iron melts in this zone.

(v) **Preheating Zone :** This zone is located above the melting zone to the charging door. Moisture and volatile matter are evaporated and the charge gets heated up in this zone.

5.3.1.3 *Raw Materials Used in Cupola*

The raw materials charged into the cupola are:

(a) pig iron

(b) scrap

(c) alloying additions to achieve the desired composition of cast iron

(d) coke as a fuel for providing the necessary heat in the cupola

(e) Fluxes like lime stone, fluorspar and soda ash to form a fluid slag.

(f) Air to provide oxygen for combustion of coke.

5.3.1.4 *Cupola Operation*

The following steps are involved in cupola operation:

(a) Preparation of Cupola

After each heat, the bottom door is opened and the contents such as unburned coke, slag and metal left in the cupola are removed. If necessary, the lining is repaired or remade. After lining repairs the bottom door is closed and duly supported using a prop. The sand bottom is then prepared with moulding sand such that is slopes (7 to 10°) towards the tap hole.

(b) Firing the Cupola

The cupola is fired by kindling wood which is placed at the bottom of the cupola. This should be done 3 hours before the molten metal is required. When the wood burns well, a small quantity of coke is dumped on it. Some more coke is placed on top of it when this coke is burning freely and thus a bed of coke is built.

The coke bed height, which is defined as the height of the coke above the tuyeres, is very important for normal working of cupola. The coke bed height should be such that all the oxygen of the air blast is consumed by the time the gases reach the top of the coke bed. Air blast pressure controls the distance of oxygen travel. Hence coke bed height is related to blast pressure as indicated below:

Blast pressure (mm of water)	Coke bed height (mm)
400	850
500	950
600	1050
700	1100
800	1150

The coke bed height is usually determined by hanging a chain or rod from the charging door. After the required coke bed height is attained and the coke is burning freely then the cupola is ready for charging.

(c) Charging the Cupola

The cupola is charged with alternate layers of metal, coke and flux. Suitable scrap is also added along with the pig iron.

(d) Opening of Air Blast

After charging the cupola, it is kept as such for about 20 minutes to 1 hour to permit the charge to preheat. After this period, the air blast is opened. The melting starts in about 8 to 10 minutes after opening the blast.

(e) Tapping and Slagging

40 to 50 minutes after opening the air blast, the first tapping can be made opening the tap hole and the slag flows from the slag hole.

(f) Closing the Cupola

When the operation is over, the blast is shut off and the prop knocked down so that the bottom door opens causing the remains to fall on the floor. Water is then sprayed on the remains and are removed from underneath the cupola.

5.3.2 Crucible Furnace

Melting in crucible furnace is one of the oldest method and is best suited for small foundries. This is because the operation of crucible furnace is intermittent and also wide range of alloys can be handled in small quantities.

In crucible furnace, the metal to be melted is put in a crucible made of clay or graphite. Then the crucible is heated using fuel. Depending upon the fuel used for heating the charge, the crucible furnaces are classified as :

(a) Pit Type Coke Fired Crucible Furnace : Generally coke fired crucible furnace is installed in a pit and so is referred to as the "pit type".

Fig. 5.26 shows the pit type coke fired crucible furnace. It consists of a cylindrical steel shell lined inside with refractory brick. The bottom is closed with a grate and the top is covered with a removable lid.

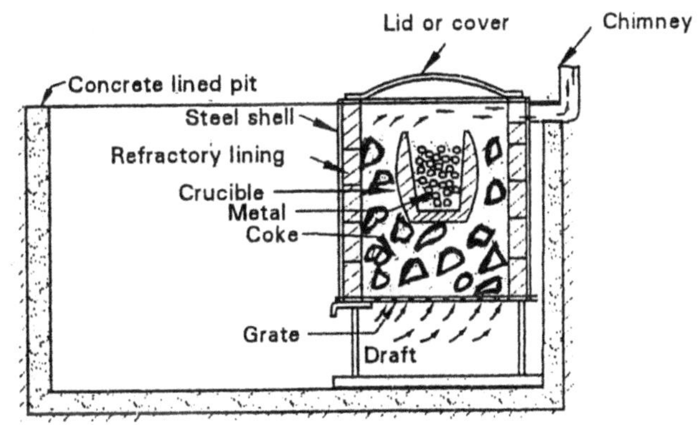

Fig. 5.26 Pit type coke fired crucible furnace.

Operation of the furnace involves kindling a deep bed of coke and allowing it to burn. Small amount of coke is then removed and the crucible is lowered into the furnace. The coke is again added surrounding the crucible on all sides. The metal is then charged in the crucible and lid is placed on the top to facilitate the chimney draft. The charge starts melting and when the metal reaches the desired temperature, the crucible is lifted from the furnace with the help of tongs and carried to the moulding section for pouring into the moulds.

The furnace is widely used for melting non ferrous metals such as brass, bronze and aluminium because of its low cost of installation, easy operation.

(b) Oil or Gas Fired Crucible Furnace :

In this furnace oil or gas is used as a fuel for heating the crucible. The furnace is a cylindrical shaped steel shell lined inside with refractory bricks. The flames produced by the combustion of oil or gas is made to sweep around the crucible and heat it uniformly (Fig. 5.27).

The advantages of this furnace are (i) no wastage of fuel, (ii) less contamination of molten metal (iii) better temperature control

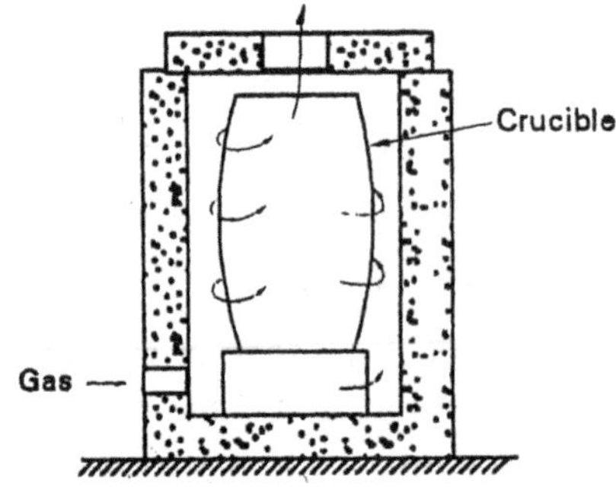

Fig. 5.27 *Gas fired crucible furnace.*

5.4 POURING MOLTEN METAL

The production of sound castings depends much upon how the molten metal enters the mould cavity through gating system.

5.4.1 Gating System

Gates are channels through which molten metal flows in order to fill a mould cavity. The term gating or gating system includes all the passage ways through which molten metal enters the mould cavity.

Fig. 5.28 shows the parts or components of a gating system. The molten metal is first poured into pouring cup or pouring basin which minimizes the splash and turbulence and allows clean metal to flow into the down sprue. Through the down sprue the molten metal enters the runners and from which it inturn passes through the gates and finally enters the mould cavity. Sprue base or sprue well serves as a cushion to the falling weight of metal and absorbs its kinetic energy.

The functions of an ideal gating system are :

    (i)   In the shortest possible time the mould cavity should be filled completely with molten metal.

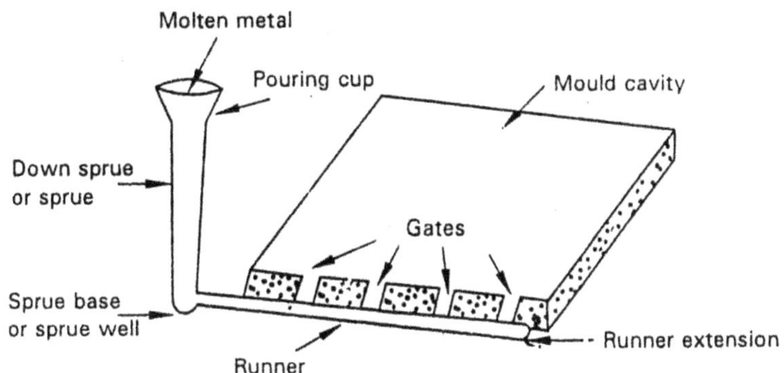

Fig. 5.28 *Components of a gating system.*

(ii) The metal should flow through the gating system with minimum turbulence and aspiration of mould gases. Excessive turbulence causes aspiration of air and formation of dross.

(iii) The metal should be introduced into the mould cavity in such a way that directional solidification (In directional solidification, the solidification will start at a farthest point from the riser and then the solidification should proceed towards the riser and the riser should be the last part to solidify) towards the riser should be easily achieved with establishing proper temperature gradients.

(iv) The casting should be produced with minimum of metal in gates and risers.

(v) By providing proper skimming action, loose sand, oxides and slag should be prevented from entering the mould cavity along with molten metal.

(vi) Entry of molten metal into the mould cavity should be regulated properly.

(vii) The erosion of the mould walls should be prevented.

For better out put and low rejects the gating system should be designed carefully. The important parts of the gating system, the pouring basin, sprue, runner and gates are discussed in detail below.

Pouring Basin

Pouring basin is a funnel shaped opening made at the top of the sprue in the cope. From the furnace, the molten metal is carried in a ladle and is poured into pouring basin and from here the molten metal flows to various parts of the mould.

Proper design of the pouring basin regulates the rate of molten metal flow, allows the molten metal flow into the sprue

Fig. 5.29 *Typical designs of pouring basins.*

smoothly and avoids turbulence. The typical designs of pouring basin are shown in Fig. 5.29.

Sprue

Sprue is the vertical passage through the cope and connects the pouring basin to the runner or the gate. The basic requirements of a sprue are as follows :

(i)   The rate of molten metal flow is determined by the sprue size. The sprue size should be small enough so that metal flows in the mould cavity at a velocity that prevents splattering and turbulence. At the same time, the size should be large enough to fill the molten metal in the mould cavity without any laps, seams or misruns. Therefore an optimum sprue size should be chosen depending upon the weight of the casting.

(ii) When the molten metal leaves the pouring basin and descends through the sprue, it is under constant gravitational acceleration. The further the metal descends in the sprue the greater it is accelerated and so the velocity of the molten metal near the bottom of the sprue is considerably greater than the velocity of the melten metal at the top of the sprue. And since volume flow rate must be same at all points in the sprue, the metal stream contracts as it descends in the sprue. This contraction creates a partial vacuum between the sprue walls and metal stream and air is aspirated into the molten metal as shown in Fig. 5.30. However, if a sprue is employed whose walls are not parallel but are tapered, the molten metal can lie firmly against the sprue walls as it descends and air aspiration is prevented.

The base of the sprue is normally made one and a half time wider and deeper than the runner which is known as sprue well. This way a stagnant pool can be created at the base of the sprue and provides a cushion to the falling weight of molten metal and absorbs its kinetic energy which minimizes erosion at this point.

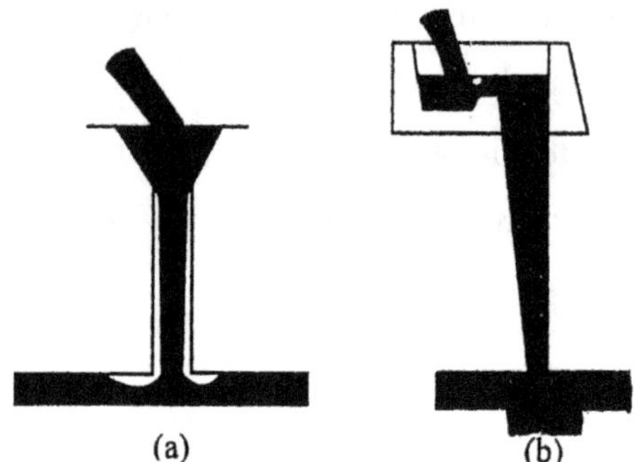

Fig. 5.30 *Effect of sprue design on metal turbulance.*

A –   Straight sprue, sharp corners : severe aspiration

B –   Tapered sprue, Rounded corners, sprue well and dam type pouring basin : Negligible aspiration.

Runner

In the production of large castings, a runner may be employed. This runner takes the molten metal from the sprue base and distribute it uniformly to several gates around the cavity.

Depending upon the shape of the casting, runner may be positioned either in the cope or in the drag part of the mould but it is generally preferred to have it in the drag.

In order to prevent the first metal poured without entering the mould cavity, runners should be tailed off (runner extension) so that any oxide, sand or other impurities washed will be carried into the ends of the runner.

5.4.1.4 *Gates*

The gate is the passage that finally leads molten metal from the runner into the mould cavity. The size and location of gates should be such that the moulds may be filled in quickly with minimum amount of cutting of the mould surfaces by the flowing metal. Gates should be located in such a way that they can be readily removed after solidification without damaging the castings.

According to their position in the mould cavity, gates may be broadly classified as :

 (i) Top gates

 (ii) Parting line gates

 (iii) Bottom gates

(i) Top Gates : In top gating, the molten metal from the pouring basin flows down directly into the mould cavity. The advantage of top gating is that since the molten metal enters the casting at the top, the hottest metal remains at the top of the casting. Due to this, proper temperature gradients favourable for directional solidification are achieved. In top gating another advantage is that the gates themselves may be made to serve as risers.

The main disadvantage of top gating is the erosion of mould due to the metal which falls directly into the mould. Therefore the mould cavity should be hard and strong enough to resist the impact of molten metal.

Various types of top gates are as follows :

(a) Slit or pencil gate

In this type of gating the sprue is made up of a series of slits which breaks up the molten metal stream into various branches thereby the head of metal weight is reduced to a large extent (Fig. 5.31). This type of gating also produce a vortex in a round sprue and hence turbulence and oxidation is avoided in the sprue.

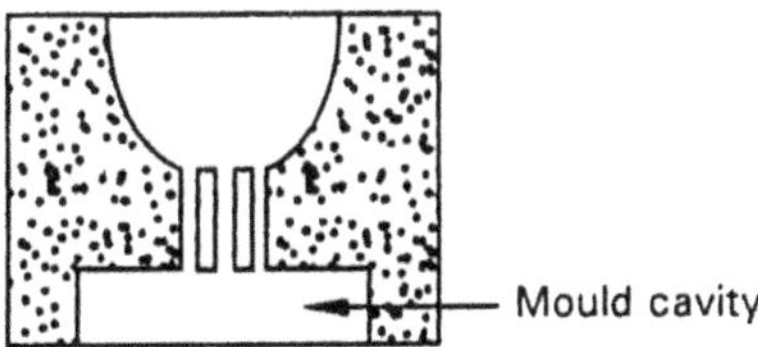

Fig. 5.31 *Slit or pencil gate.*

(b) Wedge gate

Wedge shaped gates which are called wedge gates are used in case of light castings since in light castings the weight of head is not much (Fig. 5.32).

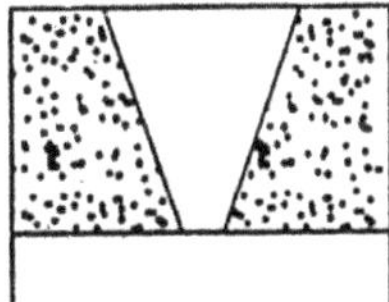

Fig. 5.32 *Wedge gate.*

(c) Top gate with a strainer

In this a strainer core or ceramic coated screen with many small holes is used in the pouring basin for better results (Fig. 5.33). The strainer restricts loose sand and dross from entering the sprue.

(ii) Parting line gates :

Parting line gates are provided at the parting surface of the mould parts i.e., cope and drag. The devices that can effectively trap any slag, dirt or sand can be used by providing the gate at the parting surface.

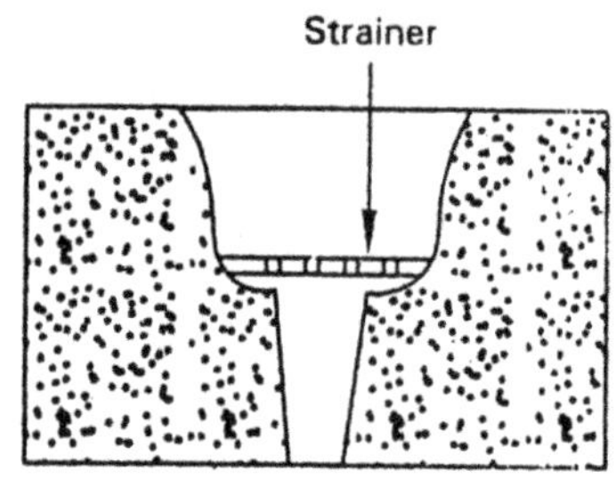

Fig. 5.33 Top gate with a strainer.

The various types of parting line gates are as follows :

(a) Skimbob

As shown in the Fig. 5.34 skimbob is a hollow recess or an enlargement in the gating. Skimbob effectively trap the slag and any foreign matter in the molten metal. Slag or any foreign matter present in the molten metal coming through the sprue will float on the surface of the molten metal and these can be trapped by the skimbob due to its curvature.

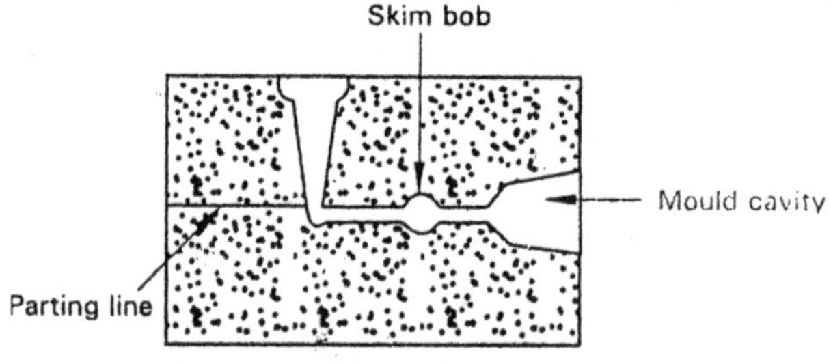

Fig. 5.34 Parting line gate with skimbob.

(b) Skimming gate

Skimming gate is a vertical passage through the cope connecting the runner (Fig. 5.35). Like skimbob, here also any foreign matter being lighter than the molten metal raises up through the skimming gate and is thus trapped.

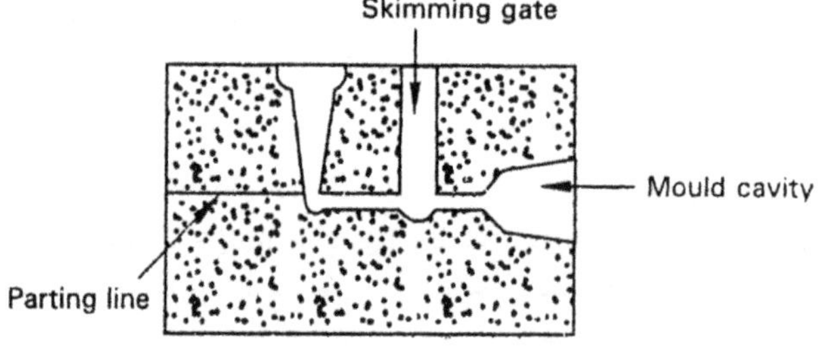

Fig. 5.35 Skimming gate.

(c) Shrink bob

Shrink bob not only acts as slag or dross collector but also acts as a metal reservoir to feed the casting as the casting shrinks during solidification (Fig. 5.36). A shrink bob is usually provided if there is a tendency for shrinkage defect to occur near the ingate.

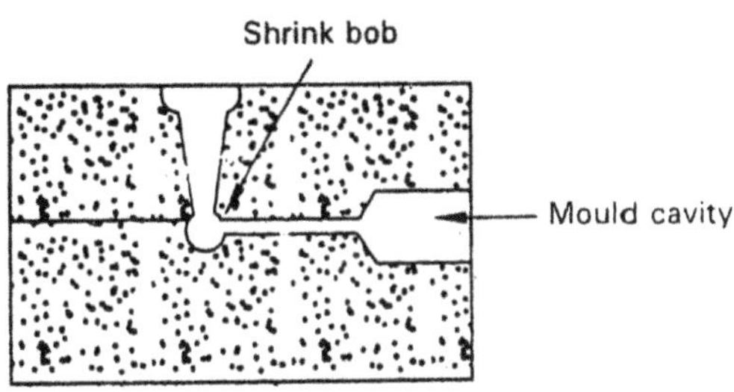

Fig. 5.36 *Parting line gate with shrink bob.*

(d) Whirl gate

The aim of whirl gate is to produce a quick rotational movement of the molten metal so that slag and any foreign matter are separated from molten metal mechanically. Here the molten metal enters the pool at an angle to the main flow and is made to spin like a whirl pool (Fig. 5.37). Due to the centrifugal force the molten metal (being heaviest) is forced outward and the foreign matter

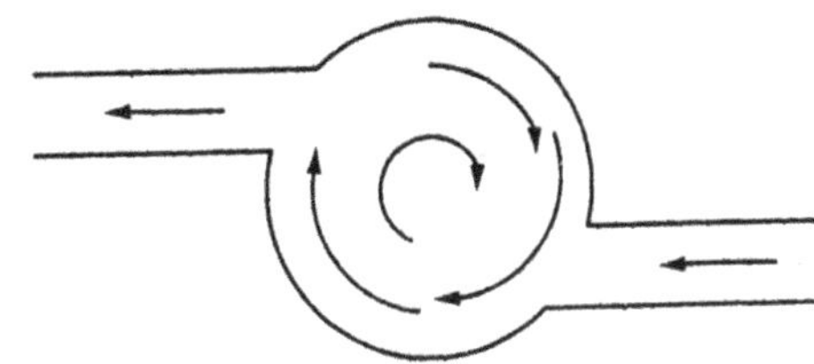

Fig. 5.37 *Whirl gate.*

(lighter) collects at the centre. The metal from the pool enters a gate which angles off in a direction opposite to that of the spinning metal in the pool. The foreign matter is collected by the riser which is provided above the pool.

(iii) Bottom gating : In bottom gating the molten metal enters the mould cavity at the bottom and the molten metal is allowed to raise gently in the mould (Fig. 5.38). The main advantage of bottom gating is that turbulence is minimum thereby the mould erosion is prevented. The disadvantages are :

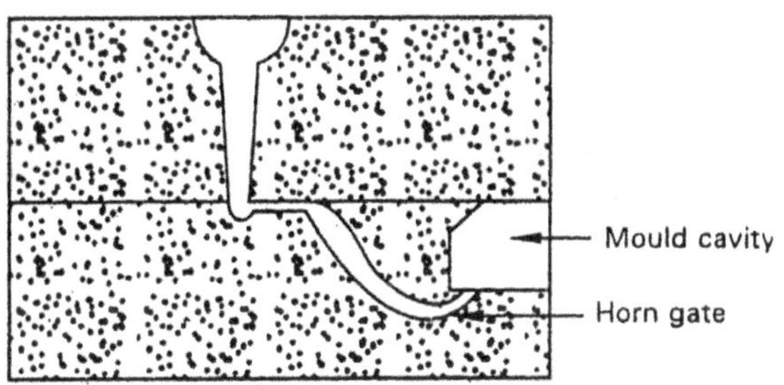

Fig. 5.38 *Horn bottom gate.*

(a) As the metal raises in the mould cavity, it continues to loose its heat and by the time it reaches the riser, it becomes much cooler. Hence directional solidification is difficult to achieve since for directional solidification the hottest metal should be in riser.

(b) It is difficult to place the riser near the gate entrance where the metal is hottest.

5.4.1.5 *Gating Ratio*

It is a term used to describe the relative cross sectional areas of the components of a gating system.

Gating ratio is defined as the ratio of sprue area to total runner area to total gate area. For example, a gating system which has a sprue of 1 sq. in., cross section, a runner of 2 sq. in., and two gates each of 1 sq. in. cross section has a gating ratio of 1:2:2.

The gating ratios are classified as pressurized and unpressurized systems.

(a) Pressurized system : In the pressurized system a back pressure is maintained on the gating system by a fluid flow restriction at the gates. This usually requires that the total gate area be not greater than the area of the sprue. For example, in systems with gating ratios of 1 : 0.75 : 0.5 or 1:2:1. Pressurized system is used for metals like steel, iron, brass etc.

Advantages :

(i)   Gating system is kept full of metal. The back pressure due to restriction at the gates tends to minimize danger of the metal pulling away from the mould walls. If back pressure is sufficient, even untapered sprues may be kept full of metal.

(ii)   When multiple gates are used flow from each of the gate if the gates are of equal area is about equal. In unpressurized systems the kinetic energy of the metal stream tends to carry it down the length of the runner and preferentially out the gate farthest from the sprue. The restricted gate areas of the pressurized system tend to minimize these kinetic effects and flow is more nearly equal from equal size gates.

(iii)   Pressurized systems are generally smaller in volume for a given metal flow rate than unpressurized ones. Thus less metal is left in the gating system and casting yield is higher.

Disadvantages :

(i)   Severe turbulence may occur at junctions and corners unless careful streamlining is employed.

(ii)   The molten metal enters the mould cavity with high velocities which results in additional severe turbulence causing air entrapment, dross formation and mould erosion.

(b) Unpressurized system : In this system, the primary restriction to the molten metal is at or very near to the sprue. The gating ratios like 1:3:3, 1:2:2 will produce an unpressurized system.

This system is adopted for light oxidizable metals like aluminium and magnesium where the turbulence is to be minimized by slowing down the rate of metal flow.

Advantages :

(i) Lower metal velocity as compared with pressurized systems.

(ii) Large cross sectional area of the runner and gates permits adequate flow rates at relatively low velocities there by reducing (a) the turbulence in the gating system and (b) spurting of the metal into the mould cavity.

Disadvantages :

(i) Careful design is required to ensure that unpressurized systems are kept filled during pouring. Often tapered sprues with sprue wells are employed. Stream lining of the junction between runner and gates is also recommended.

(ii) In this system, to obtain equal flow from multiple gates is difficult. To obtain uniform flow through all gates careful design including reduction of runner size after each gate can be used.

(iii) Casting yield is somewhat reduced since unpressurized systems require large runners and gates.

Present foundry practice favours the use of mildly pressurized systems for ferrous metals, especially for cast iron. Brasses and bronzes are gated with either pressurized or unpressurized systems depending on the individual shop and the alloy poured. The light, oxidizable metals such as aluminium and magnesium are usually gated with unpressurized systems.

5.4.2 Risering

The molten metal during solidification in the mould contracts in volume. This metal contraction takes place in three stages :

(i) *Liquid contraction* : When the molten metal cools from the pouring temperature to the temperature at which solidification just starts, liquid contraction occurs. This liquid contraction is approximately 1.6 percent by volume.

(ii) *Solidification contraction* : During the change of molten metal from liquid state to solid state, solidification contraction occurs. This contraction is usually about 3 per cent by volume.

(iii) *Solid contraction* : When the solidified metal cools from the freezing temperature to the room temperature, solid contraction occurs. This solid contraction is usually about 7.2 per cent by volume.

Of the above three contractions, solid contraction is compensated by the shrinkage allowance of the pattern maker. Liquid contraction is generally negligible. But the solidification contraction is large enough and should be taken care during designing and positioning of risers.

5.4.2.1 *Riser*

A riser is a reservoir of molten metal, attached to the casting to compensate for the shrinkage which takes place during solidification. The main requirements of a riser are :

    (i)   In order to feed the casting to soundness, the riser must be of such a volume that it has enough reservoir of molten metal.

    (ii)   The solidification time of the riser should be greater than that of the casting.

    (iii)   The riser should derive sufficient feeding pressure either by atmospheric pressure or by metallostatic pressure.

    (iv)   To achieve suitable temperature gradients for directional solidification, the riser should be the last part of a casting to solidify.

5.4.2.2 *Shape and Size of Risers*

By having minimum surface area of the riser for a given volume, the metal in the riser can be kept in the molten condition for a longer time than in the mould cavity since heat loss is minimum if the surface area for a given volume is minimum. Spherical shape riser satisfies this condition followed by cylinder and square. But moulding of spherical risers is difficult. Therefore the best shape to be employed usually is a cylinder.

As regards the height of the riser, the riser must be tall enough to make sure that any pipe formed in it does not penetrate the casting. The ratio of height to diameter usually varies from 1:1 to 3:2. As regards the diameter of the riser, it is still largely a matter of experience. For general guidance, the empirical formulae derived by Chvorinov, Caine, etc., can be used.

(a) Chvorinov's rule : It states that the freezing time of a casting i.e.,

$$T = \frac{1}{q^2} \left(\frac{V}{A} \right)^2$$

where

    q = solidification constant, depending on the composition of cast metal and the positioning of the mould cavity i.e. along a horizontal or vertical axis. For steel $q = 2.09$

    V = Volume of the casting

    A = Surface area of the casting

To determine a suitable riser diameter for a given casting, its $\left(\dfrac{V}{A}\right)^2$ ratio is computed and

a riser whose $\left(\dfrac{V}{A}\right)^2$ is a little larger (10-15 percent larger) than that of the casting is chosen.

(b) Caine's method : Caine has derived an equation for the minimum riser size effective for a given casting which is based on Chvorinov's experimental and theoretical work.

Caine defines the relative volume V of riser and casting as

$$V = \frac{\text{Volume of riser}}{\text{Volume of casting}} \qquad \ldots\ldots\ldots\ldots(1)$$

and relative freezing time X as

$$X = \frac{\text{Surface area of casting}/\text{volume of casting}}{\text{Surface area of riser}/\text{volume of riser}} \qquad \ldots\ldots\ldots\ldots(2)$$

Caine suggests a hyperbolic relationship between relative freezing time and relative volume.

i.e., $\qquad X = \dfrac{L}{V - B} + C \qquad\qquad \ldots\ldots\ldots\ldots(3)$

where $\quad B = $ relative contraction on freezing

$\qquad$ L and C = constants depending upon the metal to be cast.

The values of L, C and B for three common metals are as given in Table 5.4.

Cast metal	L	C	B
Steel	0.12	1.0	0.05
Alummium	0.10	1.08	0.06
Grey cast iron	0.33	1.0	0.03

Table 5.4

Thus for steel castings the equation (3) becomes,

$$X = \frac{0.12}{V - 0.05} + 1.0 \qquad \ldots\ldots\ldots\ldots(4)$$

Equation (4) determines the limit whether a casting will be sound or not. This equation is plotted in Fig. 5.39.

As per the Fig. 5.39, points lying to the left of the curve produce shrinkage whereas points lying on the right of the curve produce sound casting.

For a casting of given dimensions, whether or not a riser of height 'h' and diameter 'd' will be a suitable feeder can be determined as follows ;

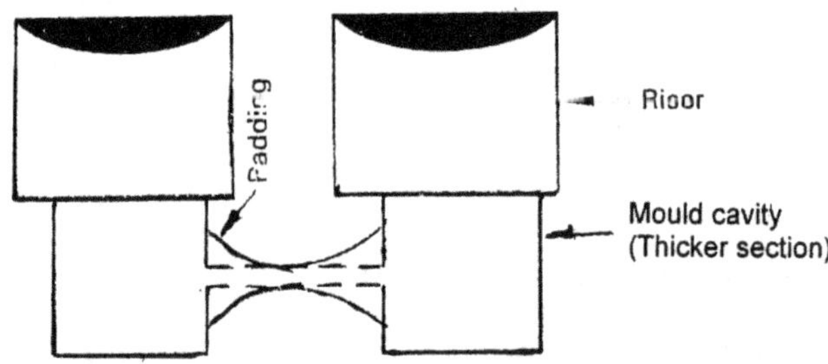

Fig. 5.39 *Caine's risering curve.*

(a) With the help of equation (1), determine the relative volume of riser and casting, say it comes out to be V_1.

(b) With the help of equation (4), determine the relative freezing time by substituting $V = V_1$ say it comes out to be X_1

(c) Now plot the point (X_1, V_1) on the graph. If this point lies on the right of the curve, the riser with given dimensions will be a successful riser. Whereas if the point lies on the left of the curve, the riser will be unsuitable and riser with another dimension should be tried.

5.4.2.3 *Improvement of Riser Efficiency*

Riser efficiency can be improved by the following methods :

5.4.2.3.1 *Padding*

Padding is usuaily used for improving casting soundness. When the casting is thinnest at points distant from the riser, freezing is complete in these regions first and hence directional solidification is difficult to achieve. In such cases, directional solidification can be achieved by padding i.e., making thinner sections gradually thicker towards the riser (Fig. 5.40). The padding is simply extra metal added to the original uniform section of the casting. After solidification, the extra metal may be removed by machining or the casting can be designed such that the tapered sections from an integral, useful parts of the ultimate assembly.

Fig. 5.40 *Padding to reduce centre line porosity*

5.4.2.3.2 Insulation Pads and Sleeves

Insulating pads and sleeves for risers are made with material which is insulating with respect to the mould material. This material should withstand the corrosive action and temperature of the molten metal.

For non ferrous castings made in sand moulds, plaster of paris is an excellent insulator. For magnesium castings asbestos sheet is sometimes used for insulating risers.

For ferrous castings, plaster of paris cannot be used because sulphur of calcium sulphate reacts readily with iron at high temperatures. Usually fire clay-saw dust sleeves are used for ferrous castings.

5.4.2.3.3 Exothermic Riser compounds

For producing directional solidification by creating heat, exothermic riser compounds (which are loose materials) are sprinkled on the top of risers. These compounds are essentially mixtures of a metal oxide and aluminium which are added to the surface of the molten metal in the riser just after pouring or to the sand of riser walls. When these compounds comes in contact with the molten metal, a chemical reaction occurs producing molten metal, heat and a slag insulator.

For example, the reaction with iron oxide and aluminium is as follows :

$$4\ Fe_2O_3 + 8\ Al \rightarrow 4\ Al_2O_3 + 8\ Fe + \text{heat } (2480\ °C)$$
$$\text{(slag)} \qquad \text{(molten metal)}$$

The slag layer (Al_2O_3) insulates the riser against heat loss to the atmosphere and the large amount of heat produced during the reaction keeps the metal in the molten state for a longer time. Similar reaction will occur with oxides of other metals such as copper, cobalt, nickel and manganese.

By employing exothermic riser compounds, the riser efficiency can be improved by about 70 percent. Hence short risers about a third the size of a normal riser can be used. Short risers are always preferred because heat is concentrated at the casting - riser neck where it is needed to prolong feeding and also suitable thermal gradients are developed for directional solidification.

5.4.2.3.3 Chills

Chills are usually employed to hasten solidification of a region of the casting which is inaccessible to a riser.

Chills are broadly classified as internal chills and external chills.

(a) Internal Chills

These are placed in the mould cavity and enter into the casting when the molten metal is poured. Hence the composition of internal chills should be same as that of the cast metal.

As shown in the Fig. 5.41, internal chills may be in the form of a nail with a large head projecting out in the mould cavity. Internal chills should be of correct size otherwise an undersized chills will not serve its purpose and an oversized chills will cause excessive chilling which prevents directional solidification.

Internal chills should be dry and clean before use and some times they are coated with tin or some other materials to prevent rusting.

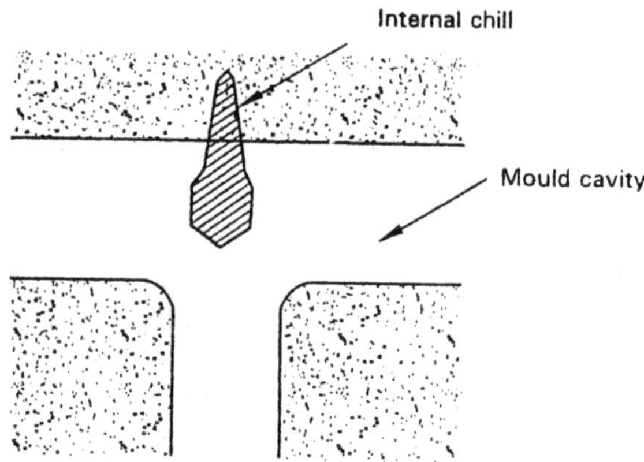

Fig. 5.41 *Typical use of an internal chill.*

(b) External Chills

External chills need not be of the same composition as that of the casting since fusion does not occur. These are usually made of metals such as cast iron, copper, steel etc. Of these copper is the best external chill due to its high thermal conductivity.

External chills are excellent for controlling cooling rates in critical regions especially for preventing the formation of hot spots at casting junctions. Some typical applications of external chills are shown in Fig. 5.42.

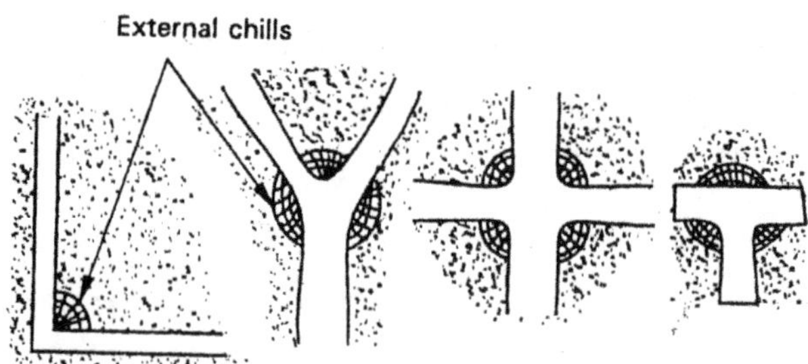

Fig. 5.42 *Use of external chills to prevent the formation of hot spots at casting junctions.*

5.4.2.4 *Types of Riser*

Risers are usually of two types viz., open riser and blind riser.

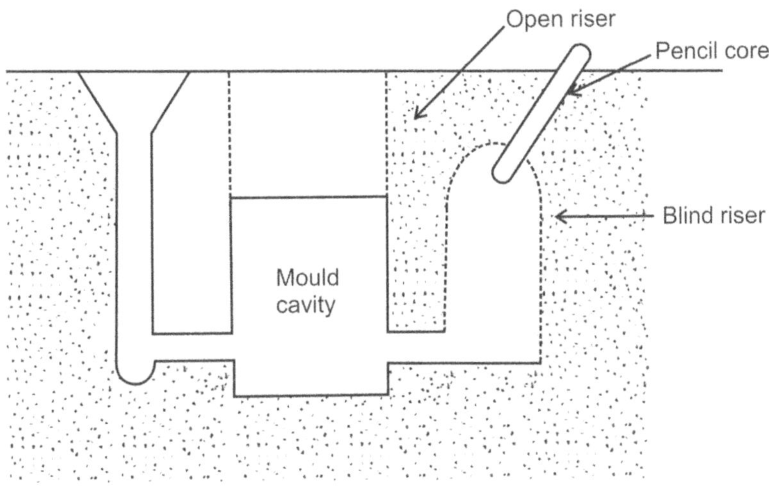

Fig. 5.43 *Position of open riser and blind riser.*

The Fig. 5.43 shows the position of open riser and blind riser.

(a) **Open riser :** The upper surface of open riser is open to the atmosphere. The open riser is usually placed on the top of the mould cavity. It derives the required feeding pressure both from atmosphere and from force of gravity. The main drawback with open riser is that when the molten metal at the top surface of the riser gets solidified then the atmospheric pressure will not act. In that case the flow of metal from the riser to casting becomes difficult. This problem can be eliminated with the help of exothermic riser compounds.

It is easy to mould an open riser than a blind riser.

(b) **Blind riser :** The blind riser is not open to atmosphere. It is completely enclosed by the sand mould. It derives the feeding pressure from force of gravity.

The blind riser is designed to have a minimum surface area per unit volume so that riser cools slowly. During solidification, the casting and the blind riser constitute a closed shell of molten metal which develops a partial vacuum due to shrinkage in the casting. This partial vacuum draws the molten metal from the riser to the casting.

Sometimes the skin of the blind riser is pierced to allow atmospheric pressure so that the blind riser can feed more efficiently the molten metal into the shrinkage regions. This piercing can be done at the top of blind riser with a core made of sand which has high permeability. Due to the permeability of the core, atmospheric air will enter into the riser and exert some pressure. Such a core is known as pencil core or Williams core and it is most commonly used in blind risers.

5.5 SOLIDIFICATION

Understanding the mechanism of solidification is a must for foundrymen. It is because proper understanding of the principles of solidification helps the foundrymen to produce sound castings without defects.

5.5.1 Principles of Solidification

The three important points which explains the principles of solidification are :

(a)   Nucleation and Growth

(b)   Heat transfer and evolution

(c)   Shrinkage

(a) Nucleation and Growth : Solidification starts by the nucleation of small grains called nuclii. Upon further drop in temperature these nuclii start growing under the existing crystallographic and thermal conditions. The composition of the alloy and the solidification rate control the size and character of these solidified grains. The growth stops after all the liquid changes to solid indicating the completion of solidification.

(b) Heat Transfer and Evolution :   The moment the molten metal enters the mould cavity, immediately heat will be extracted from the molten metal and then transferred to the mould and from the mould to the atmosphere. This amount of heat is referred as super heat (the temperature above the melting point of the metal is called super heat) which must be removed to start the solidification. When solidification starts and a portion of liquid metal becomes solid then immediately latent heat of fusion will be evolved due to the phase change of liquid to solid. And for further solidification to occur, this latent heat has to be transferred to the mould and to the atmosphere. This heat evolution and transfer will continue until all the molten metal gets solidified. And finally the solidified casting transfers its heat to the mould and then to the atmosphere as it cools to room temperature.

(c) Shrinkage : Shrinkage occurs during the three stages of cooling, that is during removal of super heat (liquid cooling), during solidification (liquid to solid cooling) and during cooling of solidified casting to room temperature. Shrinkage during liquid cooling is negligible, shrinkage during solidification should be taken care by providing proper risers and shrinkage during cooling of solidified casting should be taken care by pattern maker by providing proper shrinkage allowance to the pattern. Thus to produce sound castings, shrinkage should be properly controlled.

5.5.2 Nucleation

There are two types of nucleation :

(a) Homogeneous nucleation

(b) Hetrogeneous nucleation

5.5.2.1 *Homogeneous Nucleation*

Homogeneous nucleation occurs in hundred percent pure metals where no impurities are present at all. In this case nucleation occurs when the molten metal is under cooled or super cooled. Homogeneous nucleation of the super cooled grains depends upon two factors :

(a) Volume or bulk free energy change and

(b) Surface free energy change

(a) Volume or bulk free energy change : Free energy is a term which is frequently employed in metallurgical thermodynamics. It is denoted by the letter 'F ' and the change in free energy is denoted by 'ΔF'. If the value of ΔF is negative then the process is spontaneous, if ΔF is positive then the process is non spontaneous and if ΔF is zero then the process is at equilibrium.

Let ΔF_v be the free energy change involved in the creation of spherical shape nucleus of unit volume. The phase change from molten metal to solid phase always accompanies free energy decrease. And therefore ΔF_v is always negative. This decrease in free energy contributes to the stability of the new phase i.e., solid phase.

Since ΔF_v is the bulk free energy change associated with the creation of a unit volume of the spherically shaped nucleus then the total bulk free energy change associated with the formation of a nucleus of radius 'r' will be

Total bulk free energy change $= \Delta F_v \times$ volume of nucleus

$$= \frac{4}{3} \pi r^3 . \Delta F_v$$

(b) Surface free energy change : Nuclii formed in the melt have some surface area. Solid-liquid phases possess a surface between the two. Such a surface has a positive free energy per unit area associated with it.

Let ΔF_s be the free energy change associated with a unit area of the surface created between the solid nuleus and liquid. And hence the total free energy change associated with a solid spherical nucleus of radius 'r' will be:

Total surface free energy change $= \Delta F_s \times$ surface area of the nucleus

$$= 4\pi r^2 . \Delta F_s$$

$\therefore$ The net free energy change ΔF is given as,

ΔF = Total bulk free energy change + Total surface free energy change

$$= -\frac{4}{3} \pi r^3 . \Delta F_v + 4\pi r^2 . \Delta F_s \qquad \qquad .....(5.1)$$

The negative sign with the total volume free energy change is due to the fact that ΔF_v is always negative.

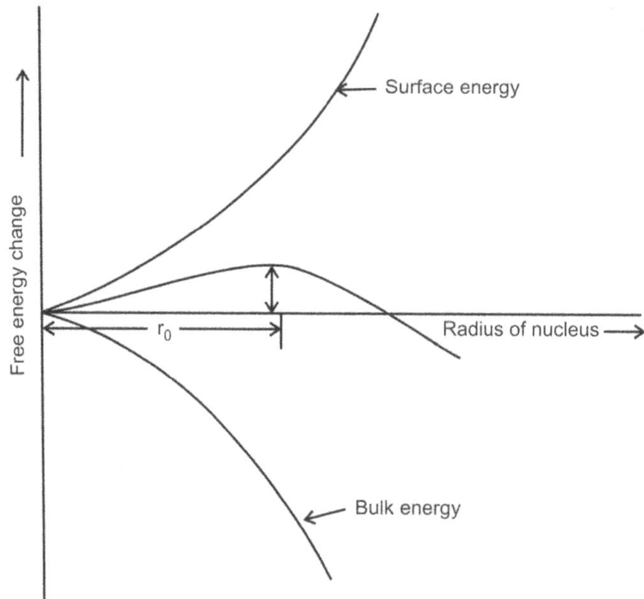

Fig. 5.44 *Free energy change as a function of radius of nucleus.*

Fig. 5.44 is a plot of the eq. (5.1). The curve of bulk free energy change versus radius of nucleus will lie on the negative side of the free energy axis since the bulk free energy (ΔF_v) is always negative. And the curve of surface free energy change versus radius of nucleus will lie on the positive side of the free energy axis since the surface free energy change is always positive.

At small radii, the surface free energy change is larger than the bulk free energy change and hence the net free energy at small radii is positive. However this situation changes as the radius grows in size because at large values of 'r' the bulk free energy change is larger than the surface free energy change and hence the net free energy at large radii becomes negative. Therefore the curve of net free energy change versus radius of nucleus, raises to a maximum at a critical radius of the nucleus and then falls. A nucleus will be stable only when its radius is not smaller than 'r_0'(the critical radius), because for any smaller value of 'r', the net free energy change increases when the nucleus begins to grow. Therefore radii smaller than the critical radius 'r_0' will be unstable.

This critical radius 'r_0' may be found by differentiating the eq. (5.1) with respect to 'r' and then maximizing by equating it to zero.

Thus,

$$\frac{d}{dr}\,(\,\Delta F\,) = \frac{d}{dr}\,(-\frac{4}{3}\,\pi r^3\,.\,\Delta F_v\,) + \frac{d}{dr}\,(\,4\,\pi r^2\,.\,\Delta F_s\,) = 0$$

$$= -3\,.\,\frac{4}{3}\,\pi r^2\,.\,\Delta F_v + 8\,\pi r\,.\,\Delta F_s = 0$$

or $\qquad\qquad r_0 = \dfrac{2\Delta F_s}{\Delta F_v}$

5.5.2.2 *Hetrogeneous Nucleation*

Practically all metals will solidify with heterogeneous nucleation since hundred percent purity metals are not available.

Hetrogeneous nucleation is easy because the formation of critical size nucleus is enhanced by certain substances called as nucleation catalysts or nucleants. All practical metals possess these nucleants. The nucleants can be solid particles suspended in the liquid metal, the surface of the mould or the solid films such as oxide on the surface of the liquid. These foreign particles help in nucleation and thereby reduce the amount of super cooling needed to accelerate the nucleation. Hence heterogeneous nucleation is easier to occur than homogeneous nucleation.

5.5.3 Growth

After the formation of the stable nuclii, growth of the nuclii starts. Stable nuclii will grow by acquiring the atoms from the liquid. The rate of growth depends upon the amount of super cooling. Usually in the beginning of solidification, nucleation is predominant and growth occurs subsequently. Since in the early stages nucleation is predominant, the first layer of solid metal that is formed at the mould wall consists of fine equiaxed grains.

Growth is normally controlled by the rate of heat transfer from the solidifying casting and this growth occurs in a direction opposite to the heat flow. That means the growth proceeds from the mould wall to the centre of the casting progressively. This solidification is called as progressive solidification where the solidification starts at the mould wall and proceeds progressively towards the centre of the casting from all directions.

Apart from the direction of heat flow, growth is also dependent on the crystallographic direction. Hence only those nuclii which are favourably oriented towards the direction of heat flow and the crystallographic direction can grow towards the centre of the casting and other nuclii which are not oriented favourably will not grow and peel off into the liquid. The net effect will be to create a zone of columnnar grains next to the fine equiaxed grains which are at the mould wall.

5.5.4 Solidification of a Pure Metal

Usually in the mould cavity, the heat extraction is maximum at the mould wall. Therefore, when a pure molten metal is poured into a mould cavity and allowed to solidify then the portion of the molten metal which first reaches the freezing temperature is at the mould wall where heat extraction is maximum. This chilling action of the mould wall causes the formation of a thin skin or shell of solid metal consisting of fine equiaxed grains in the vicinity of mould wall. Then this skin starts growing progressively inward towards the centre upon further heat extraction. And during this solidification process, the general decrease in temperature is interrupted as the first solid metal releases its latent heat of fusion. This increases the overall time required for solidification because for further solidification to occur this latent heat has to be extracted and normal solidification will occur only after all the latent heat of fusion is removed.

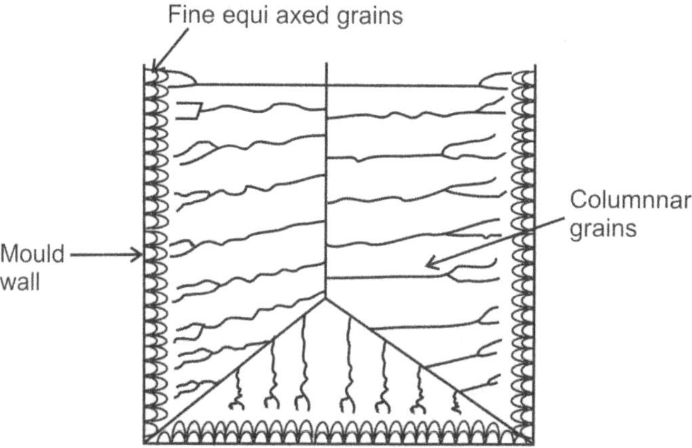

Fig. 5.45 *Showing fine equi-axed grains and columnar grains – typical of pure metals.*

Fig. 5.45 shows fine equiaxed grains at the mould wall and columnar grains extending towards the centre. This is the typical solidification mode of pure metals – fine equiaxed grains will be formed at the mould wall and then some of these grains which are favourably oriented towards the direction of heat flow and also towards crystallographic direction will grow and forms columnar grains which extend towards the centre.

In pure metals, grain growth may occur dendritically if the molten metal is super cooled. Dendritic type of growth represents only about 10 percent of the total freezing process of pure metals whereas it is a normal phenomenon incase of freezing of alloys.

5.5.5 Solidification of Alloys Where No Eutectic Occurs

In these alloys freezing occurs over a temperature range. And due to this temperature range, there is no line of demarcation between liquid and solid as in case of pure metals which freezes at constant temperature. In this case similar to pure metals, the first metal to freeze is at the mould surface and freezing progresses inward towards the centre. But the difference lies in the fact that though the start of freezing progresses inward, the metal dendrites extending into the metal leave behind islands of liquid metal. These islands freeze only on further heat extraction . Therefore when such alloys solidifies, the start of freezing proceeds in a wave like manner towards the centre of casting followed by end of freezing which is lagging behind but also moving towards the centre of casting.

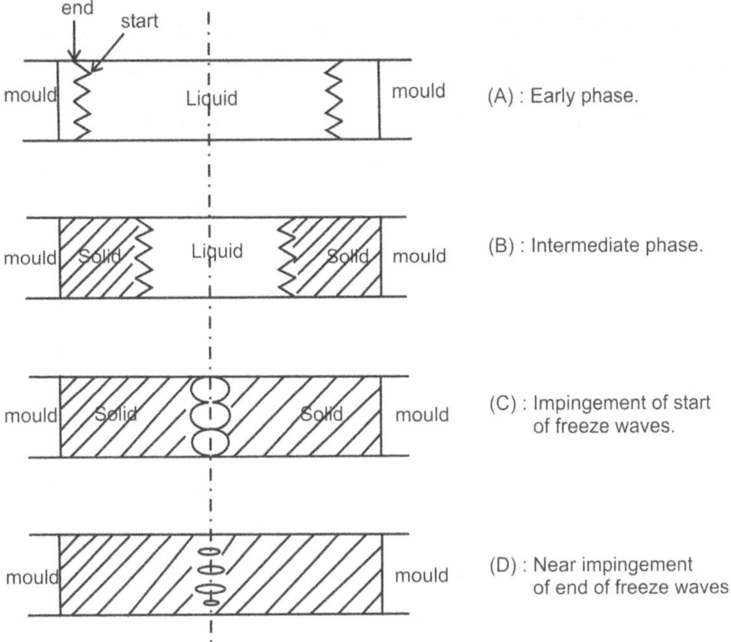

Figure 5.46 solidification mode of alloys where no eutectic occurs.

Fig. 5.46 shows the solidification made of alloys where no eutectic occurs.

Figure 5.46 (A) shows the early phase where the start of freezing and end of freezing waves starting to move into the liquid from the mould wall towards the centre of the casting.

Figure 5.46 (B) shows the early phase where the start and end of freezing waves advances towards the centre.

Figure 5.46 (C) shows the impingement of start of freezing waves indicating near completion of solidification.

Figure 5.46 (D) shows near impingement of end of freezing waves indicating the completion of solidification.

5.5.6 Solidification of Alloys With Eutectic

Solidification in these alloys occur in two stages:

(a) the dendritic growth of primary phase and

(b) the final solidification of the liquid as eutectic mixture

Therefore in this case, the first stage of solidification is the formation of dendrites of the primary phase and the second stage is the final solidification of the remaining liquid as eutectic mixture.

In these alloys the solidification mode is that first dendrite start wave will move inward towards the centre followed by dendrite end wave which also will move inward towards the centre. Then after near completion of dendrite end wave travel, the eutectic freezing starts and the final solidification of the liquid occurs as a eutectic mixture.

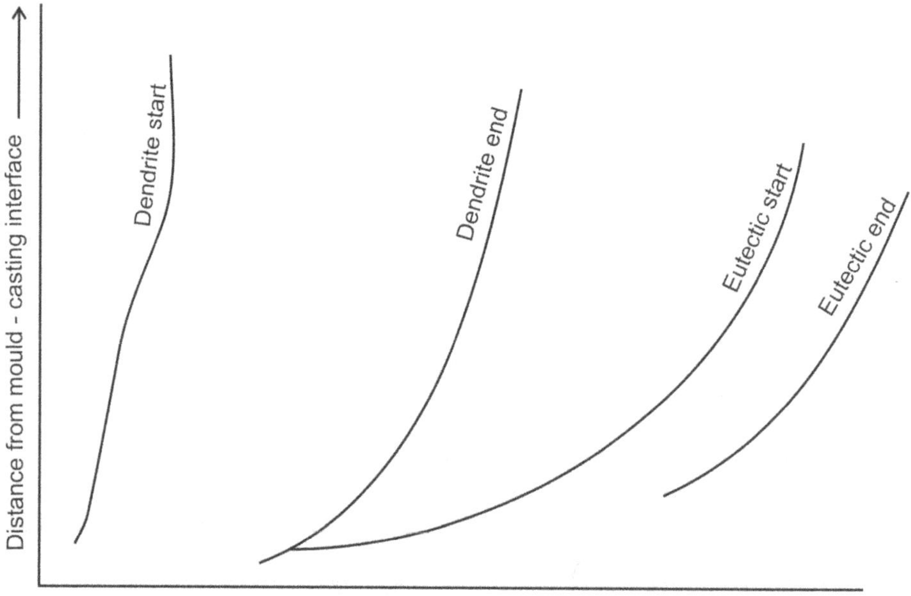

Fig. 5.47 *Solidification mode of alloys with eutectic.*

Fig. 5.47 shows for dendrite freezing the start and end waves are widely separated where as for eutectic freezing, the separation between the start and end waves is narrow.

5.6 FLUIDITY

Physicists define fluidity as a reciprocal of viscosity. But for foundryman fluidity is not the reciprocal of viscosity, it is the mould filling ability of the molten metal. Hence to produce sound castings, fluidity is a very important aspect for foundrymen. Fluidity is measured by pouring the molten metal in a standard mould. This standard mould is usually in the form of a spiral and the length of travel of molten metal in the spiral after pouring the molten metal serves as a measure of fluidity.

5.6.1 Factors Affecting Fluidity

The factors that affect the fluidity of a metal or alloy are as follows :

   (a) Viscosity

   (b) Oxide films

   (c) Gas solubility effects

(d) Surface tension

(e) Inclusions

(f) Method of solidification and growth

(g) The amount of super heating

(h) Pouring rate and head of metal used

(i) The type of mould material

All these factors affect the mould filling ability (or) fluidity of molten metal. The detailed discussion on these factors is as follows :

(a) **Viscosity :** It is difficult to measure metallic viscosity. The conditions under which the tests carried out are rarely reproducible in practice. And also in general the change in viscosity with temperature is not appreciable. Hence viscosity may not be considered as a major factor in metal filling ability.

(b) **Oxide films :** Oxides are of three main types.

Type (i) : These oxides form to a small extent and enter into solution in the metal.

Type (ii) : These oxides form a comparatively loose film on the metal surface.

Type (iii) : These oxides form a tough adeherent skin on the metal surface.

Oxides of type (i) are comparatively harmless in the quantities normally present and negligible effect on fluidity. Type (i) oxides include FeO, NiO and Cu_2O.

Oxides of type (ii) are quite harmful as compared to those of type (i). For example, take a 60:40 brass. When this alloy is in molten condition, a thin and continuous film of ZnO will be formed on the surface.

If this thin film is removed by skimming then it is immediately replaced by further oxide formations. Therefore, incase of this alloy a thin oxide film will always be present on the molten surface. The presence of this thin loose oxide film will reduce the fluidity of this alloy.

Oxides of type (iii) affects fluidity to a great extent than that of type (ii). Type (iii) oxides form a quite tough and adherent skin on the molten metal surface. For example, incase of ferrous and non ferrous alloys which contain aluminium as alloying element, a thin adherent alumina (Al_2O_3) film will be formed. This thin adherent alumina film will be an obstacle to the mould filling ability of the molten metal.

(c) **Gas solubility effects :** During pouring of molten metal into the mould cavity, some amount of gases will be evolved either from the molten metal or caused by metal-mould reaction (evolution of steam). These gases may form a barrier between the molten metal stream and the mould wall and causes the fluidity of the molten metal to decrease. Therefore fluidity may be indirectly linked with permeability of the mould.

(d) Surface tension : Bircum Shaw has measured the surface tension of several pure metals such as iron, copper, silver, zinc, tin, lead and antimony and stated that the higher the melting point of metals the higher the surface tension in most cases. It was also found that the surface tension generally decreases as any given metal is superheated.

But this surface tension variables are difficult to measure in case of alloys due to the presence of oxide films. Since almost all foundries are centred around alloys rather than pure metals, it is difficult to obtain much useful information on the effect of surface tension regarding fluidity.

(e) Inclusions : In molten metals, the inclusions held in suspension tend to reduce the fluidity. But in most of the cases, the inclusions will not be present in sufficient quantity to have a noticeable effect on fluidity.

(f) Method of solidification and growth : Regarding the effect of this factor, Portvein and Bastein examined number of metals and alloys and made two important conclusions :

   (i)   For the same super heat, fluidity is greatest for pure metals or eutectics and least for saturated solid solutions.

   (ii)   Fluidity is largely dependent upon the types of crystal formation and tends to be greatest when the crystals formed are of non dendritic type.

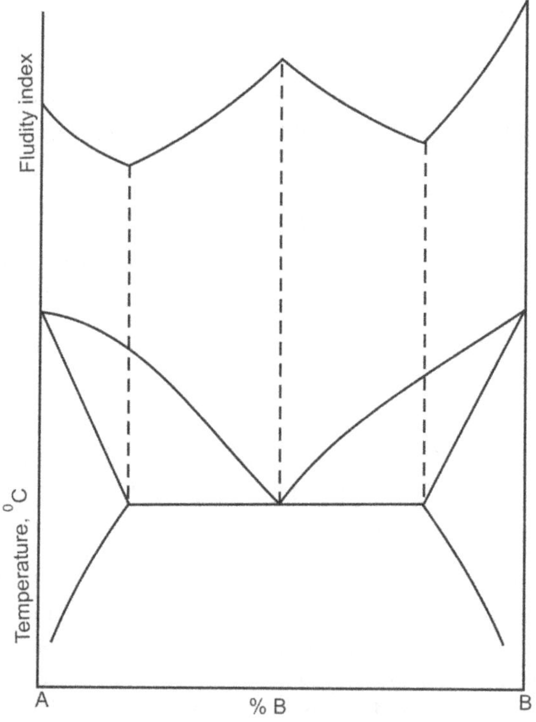

***Fig. 5.48** Typical fluidity index for a binary alloy system.*

Fig. 5.48 shows a typical fluidity index for a binary alloy system which indicates greater fluidity for pure metals, eutectics and lower fluidity for saturated solid solutions.

(g) Super heating : To study the effect of super heating on fluidity, a series of spirals are poured at different pouring temperatures by keeping the alloy composition same for various alloys.

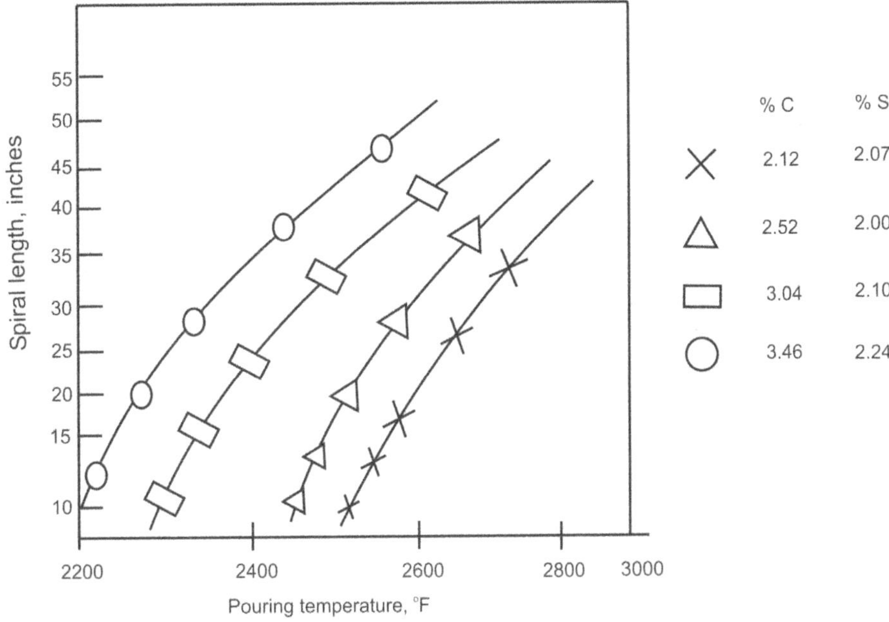

Fig. 5.49 *Effect super heat on fluidity for different cast irons.*

The curves in the Fig. 5.49 shows a linear variation for spiral length versus pouring temperature particularly at shorter spiral lengths.

The curves also show that a low carbon cast iron has low fluidity and gives the information that practically any fluidity can be obtained by increasing the pouring temperature. But the problem is the mould materials can not tolerate the high pouring temperatures.

(h) Pouring rate and head of metal used : Consistent pouring of molten metal is difficult and hence pouring rate is capable of variation which mainly depends upon the ladle design and the operation (pourer). It is common practice to pour the alloy at the maximum possible rate to achieve the better fluidity.

The head of the metal in the pouring basin is equally important to achieve the good fluidity. Maintenance of sufficient head of the metal results in better fluidity for the molten metal.

(i) **Mould material :** Since the thermal conductivity of the mould wall influences the cooling rate of the molten metal, the mould material plays an important role in mould filling ability of the molten metal. Preheating the mould which delays the solidification of molten metal improves the fluidity. For example, thin stainless steel casting can be made by pre heating the mould in excess of 800 °C.

Mould surface finish also plays an important role in mould filling ability or fluidity. Generally moulds finished to close limits will allow the molten metal to flow freely thereby improving fluidity. Better mould surface finish gives better fluidity for the molten metal.

Similarly mould dressings, which improve the mould surface finish, play an important role to give better fluidity to the flowing molten metal.

5.6.2 Fluidity Measurement

Though there are number of methods, here the two methods used for fluidity measurement are discussed.

(a) Vacuum suction method :

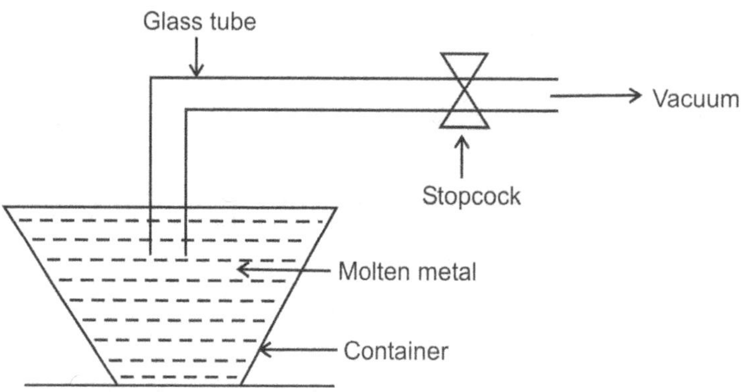

Fig. 5.50 Equipment for vacuum suction method.

In this method, molten metal is taken in a container and then a glass tube is dipped into the molten metal. A stop cock connecting the upper end of the tube to a vacuum tank is then opened (Fig. 5.50). The molten metal immediately flows up in the glass tube under the influence of the pressure head created by the difference between atmospheric and tank pressures. The length of the metal column is taken as an index of the fluidity.

This method has two advantages :

    (i)   It permits direct observation (due to glass tube) during metal flow and

    (ii)   It is somewhat simpler than pouring a sand cast spiral.

A possible disadvantage is that the thermal gradients and the nucleation conditions produced by glass mould wall are different from those arising in a sand mould.

Data obtained by this method is confirmed the greater fluidity of pure metals and eutectics.

This method is usually used for lower melting point metals such as aluminium, zinc, tin and magnesium alloys.

(b) Sand mould fluidity spiral method :

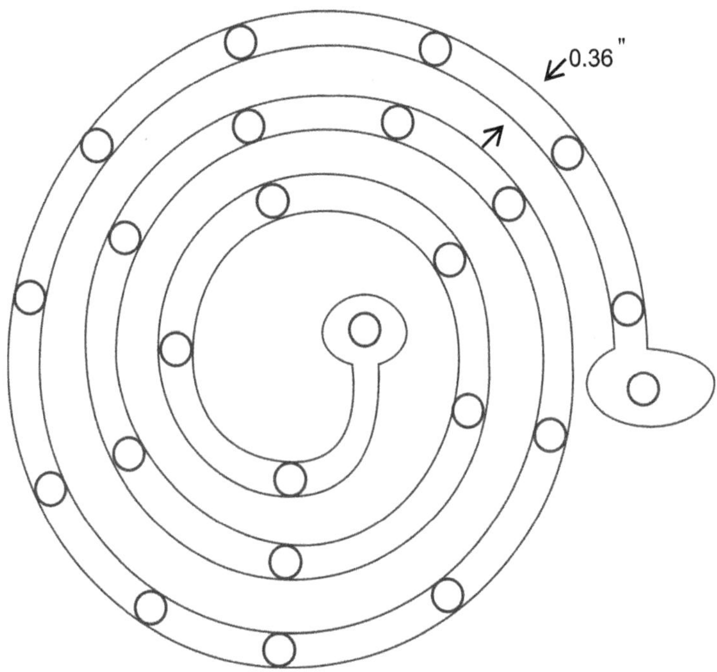

Fig. 5.51 *Fluidity spiral.*

In the early designs of the fluidity spiral, long thin sections were cast. But the data obtained is not reproducible for various reasons. Finally over the past 50 years a compact and reproducible design has been developed (Fig. 5.51). This design consists of a spiral with 5.5 inches long and this can be completely filled only under conditions of very high fluidity.

When the molten metal is poured into this spiral, the length of the partially filled spiral provides an index of relative fluidity.

This method confirms best fluidity for pure metals and eutectics and poorest fluidity for alloys having more solidification range.

5.7 CLEANING OF CASTINGS

After the molten metal has poured into the mould, it is allowed to solidify in the mould itself. Once the molten metal has solidified in the mould completely, the sand mould is to be broken to remove the casting from the mould. The removal of casting from the mould is done only when the casting is cooled sufficiently to a lower temperature otherwise if the hot casting is exposed to air uneven cooling occurs which give rise to distortion and cracks. Hence moulds should be broken at a temperature when no transformation occurs.

The moulds are broken either manually on the pouring floor or transferred to a separate shake out section. Shake out is also carried out either manually or mechanically. In mechanical shake out machine a screen is vibrated which produces jarring action, causing the separation of the moulding boxes, casting and sand. The sand falls through the screen either into a pit or on a belt conveyor and the sand is returned to the reconditioning unit. The casting and moulding boxes remain on the screen. The mould boxes are removed and returned to the moulding section whereas the casting is placed for cleaning in cleaning section.

When the casting is removed from the mould, it can not be used directly because sprue, gates, risers etc., are attached to it and also lot of sand and fused particles are clinging to the to the surface of the casting. Hence cleaning of the castings which is called as fettling involves the following three operations:

    (a) Removal of dry sand cores

    (b) Removal of extra parts like sprues, gates, risers, fins etc.

    (c) Cleaning of the casting surface.

5.7.1 Removal of Dry Sand Cores

Cores are removed simply by pushing with a bar or some times by a core vibrator. For the removal of cores from large castings hydro blasting or Electric hydraulic decoring, is effective. In Electric hydraulic decoring, hydraulic shock waves are created in a tank of water in which the castings are immersed. These shock waves, produced by high voltage electrical discharge across two electrodes submerged in the water are sufficient to loosen hard and intricate cores.

5.7.2 Removal of Extra Parts

Extra parts such as sprues, gates, risers etc. can be removed by the following methods :

    (a) Flogging

    (b) Chipping

(c) Shearing

(d) Sawing

(e) Flame cutting

(f) Plasma arc cutting

(g) Abrasive wheel cutting

Flogging

Flogging with a hammer is a means of removing extra parts by impact for brittle castings. Usually with hammers the extra parts will be knocked off. But in this case there is a danger of extending the break into casting proper. This can be avoided by notching the extra part ahead of the casting. This method is most suitable for brittle castings like grey iron and white iron.

Chipping

Chipping of extra parts is carried with the help of hand or pneumatic chisels. This method is most suitable for hard and brittle castings such as grey and white cast irons. Chipping by chisels produces a rough surface and is not suitable for ductile castings.

Shearing

Sprue cutter involves the principle of shearing in which the extra parts are sheared off ahead of the casting. Ductile metal castings such as steel, copper, brass, aluminium can be conveniently handled by this method provided the size of the casting is not too large to shear

Sawing

Band sawing is more common for both ferrous and non ferrous castings to remove extra parts since it gives faster rate of production than ordinary saws. Sawing is extensively used to remove extra parts for non ferrous castings.

Flame cutting

Flame cutting using oxyacetylene gas is generally used for large sized steel castings where the extra parts are of bigger size. The main advantage of this method is that the ease of moving the cutting torch from one portion to another while removing extra parts.

Plasma arc cutting

Plasma is produced by passing a stream of suitable gas through an electric arc discharge. The stream of plasma coming out of nozzle of a torch at a very high temperature (usually $10000 - 150000\ °C$) with a high velocity. This high temperature of the plasma is utilised in cutting extra parts. This plasma arc cutting is now used increasingly to cut extra parts in order to eliminate manual operations and to make the work fast, clean and accurate. Nowadays a programmable robot is employed in plasma arc cutting.

Abrasive wheel cutting

In this method, abrasive cut off wheels are employed for removing extra parts. Abrasive wheel cutting is used for all alloy castings but is specially designed for hard and difficult to saw castings.

5.7.3 Cleaning of Casting Surfaces

Casting surfaces are cleaned by the following methods :

 (a) Wire brushing

 (b) Tumbling

 (c) Sand blasting

 (d) Shot blasting

 (e) Hydro blasting

 (f) Grinding

 (g) Pickling

Wire brushing

Wire brushing is employed to remove sand particles from the casting surfaces. Usually rectangular, circular or cup shaped brushes are commonly used for this purpose.

Tumbling

In this method, the castings to be cleaned are placed in a large steel shell or barrel which contains small, hard, star shaped pieces of cast iron. The barrel is mounted on horizontal trunnions and can be rotated at a speed of 25 -50 rpm. When the barrel is rotated cleaning of casting surfaces occur due to the rubbing action of castings and cast iron pieces and also due to the rubbing action of castings against each other. Twenty minutes to an hour of tumbling is used for grey and malleable iron castings. In tumbling one precaution to be taken is that heavy castings cannot be charged with small fragile castings. Usually small castings which are not fragile are best suited for tumbling.

Sand blasting

In sand blasting a stream of sand is directed at a high velocity against the casting surfaces. Sand blasting is usually carried out using coarse sand (6-30 mesh size) and high velocity air as the carrying medium. One disadvantage with sand blasting is that sand particles will break down in size as they impact on hard casting surfaces and also silica dust formed during sand blasting contaminates the atmosphere. Sand blasting is useful for both fragile and large sized castings.

Shot blasting

In shot blasting, the metallic particles (shots) are thrown on casting surface by centrifugal force from a rapidly rotating impeller wheel. The adhering sand and scale on the surface of castings will be removed due to the impact of abrasive metallic particles. In this case, the cleaning rate is controlled by the size of the shots. If the size is larger, the impacts on the castings surface are insufficient and if small size particles are there, the impacts will be more but the energy of each impact may be insufficient to remove adhering sand particles from the castings surface. Therefore an optimum size of shots should be choosen. Shots may be made of white cast iron, steel etc.

Hydro blasting

In hydro blasting, at a high pressure a fine stream of water and sand is directed against the casting surface. This method cleans the casting surfaces very thoroughly and at a rapid rate. The dust problem which exists in sand blasting is eliminated completely in this method.

Grinding

Grinding of castings is carried to remove excess metal. Various types of grinders are stand grinders, portable grinders and swing frame grinders. For large sized castings, swing frame grinders are used where as stand grinders are used for small sized castings.

Pickling

Pickling means cleaning of casting surfaces with acid treatment. In this method, the castings are suspended in a tank containing equal part of hydrofluoric acid and sulphuric acid or only sulphuric acid for about four hours. The castings are then removed from the tank, washed with water and then immersed in a neutralizing tank, containing 10% solution of washing soda in order to neutralize the remaining acid on the casting surface. This method effectively removes adhering sand particles or scales and produces clean casting surfaces at a cheaper cost.

5.8 INSPECTION

Cleaning follows inspection of castings. The inspection is carried to detect defective castings. Many types of defects occur during casting due to the irregularities in the casting process. In this section the occurrence of various casting defects is discussed followed by inspection of casting to detect these defects.

5.8.1 Casting Defects

The presence of defects in casting is not accidental but they arise if some step in the manufacturing process is not properly controlled. Some of the defects may be tolerated and some of them may be eliminated by good moulding practice or repaired by welding. However, the producer, designer, engineer and customer should have thorough knowledge about the defects occuring in castings and their origin (causes) and control (remedies)

The various defects that occur in sand castings are as follows:

(i)   Defects due to improper moulding materials and core making materials.

(ii)   Defects due to improper moulding

(iii)   Defects due to molten metal

(iv)   Defects occuring during solidification

(v)   Metallurgical defects

Defects due to improper moulding materials and core making materials

The defects occuring due to improper moulding materials and core making materials are blow holes, scab. drop, metal penetration and rough surfaces, runout, fusion, rat tails and buckles, swell etc

(a) Blow holes

Blow holes are smooth and round holes present either inside the casting or on the surface of the casting. If the holes are present on the surface of the casting, they are called as open blows and if the holes are present inside the casting, they are called as blow holes.

These are caused due to the generation of steam and gases during pouring of metal and the inability of the mould to disperse them effectively. The reason for steam generation is due to the presence of excess moisture in moulding sand and cores. And the reason for gas generation is due to the presence of gas producing ingredients in the mould or core sands. The inability of the mould to disperse steam and gases effectively arises due to low permeability of the mould, and insufficient venting of the mould. Low permeability of mould arises due to extra ramming, too fine sand grains and higher clay content in the sand. By properly controlling the above factors blow holes can be avoided.

(b) Scab

Scabs are a sort of projections on the casting which occur when a portion of the mould face lifts and metal flows underneath in a thin layer. That is as shown in Fig. 5.52, the liquid metal penetrates behind the surface layer of sand.

Scabs occur in castings due to hard and uneven ramming, excess moisture in sand and cores, low permeability of moulding sand.

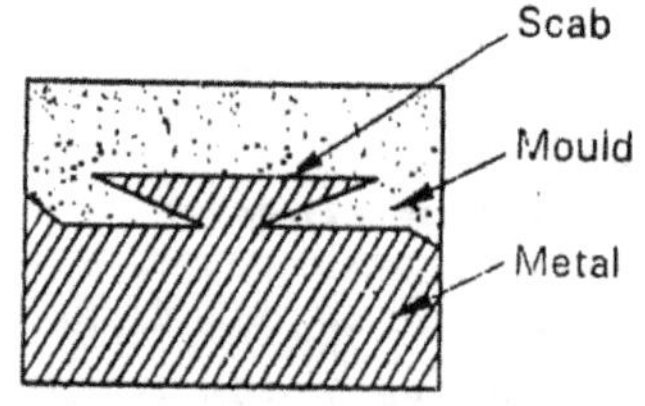

Fig. 5.52 Scab defect.

(c) Drop

Drop is an irregular deformation of casting which occurs when the upper (cope) surface of the mould cracks and pieces of sand drops into the molten metal. Drop occurs due to low green strength and improper ramming of the cope flask. By proper ramming and the use of good binder can eliminate this defect.

(d) Metal penetration and Rough surfaces

Rough casting surfaces are obtained when the molten metal fills the gaps between the sand grains. This is caused due to high permeability of moulding sand using sand having coarse grain size and soft ramming and also due to higher pouring temperatures. This defect may be eliminated by controlling the above factors.

(e) Run out

Run out results in incomplete casting and this occurs when the molten metal leaks out of the mould cavity. Run out occurs due to faulty moulding box equipment and improper moulding.

(f) Fusion

Fusion of sand grains on the casting surfaces produces a rough glassy appearance. Fusion occurs due to low refractoriness of moulding sand and too high pouring temperature. This defect may be eliminated by controlling these two factors.

(g) Rat Tails and Buckles

Rat tail is in the form of irregular projections on the surface of a castings. This is called rat tail because these irregular projections appear as if a rat has dragged its tail over the surface of the mould. Rat tail occurs due to the compression failure of the skin of the mould cavity because of the high pouring temperature.

Severe rat tails are called buckles

The occurrence of rat tails and buckles are due to poor expansion characteristics of moulding sand, lack of combustible additives in the moulding sand and too high pouring temperature. By controlling these factors the rat tails and buckles may be eliminated.

(h) Swell

Swell is an enlargement of the mould cavity due to molten metal pressure which results in localised or overall enlargement of the casting.

Swell occurs due to insufficient ramming of the moulding sand. Insufficient weighing of the mould boxes during pouring may also cause the cope to lift giving a swell. By correcting these factors swell may be eliminated.

Defects due to improper Moulding

The defects occuring due to improper moulding are mismatch or mould shifts, fins, crush, etc.

(a) Mismatch or Mould shifts

A shift results in a mismatch of the sections of a casting usually at the parting line. Fig. 5.53 shows the mismatching of top and bottom parts of the casting at the parting line.

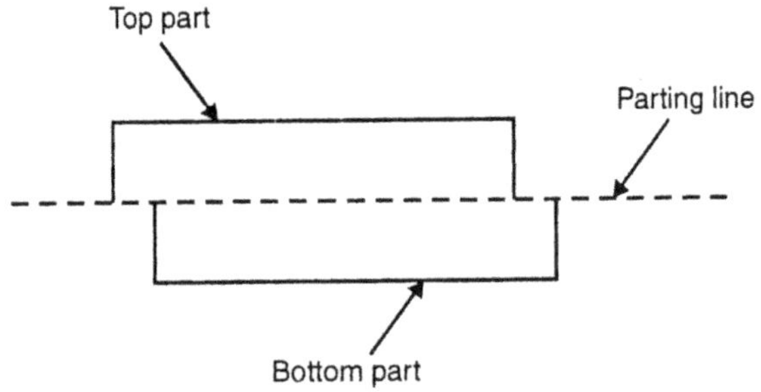

Fig. 5.53 Mismatch.

Mould shift occurs due to not locating the pins properly on pattern plate when using half patterns and also due to misalignment of moulding flasks.

Shift may be eliminated by ensuring proper alignment of the moulding flasks, correct mounting of patterns on pattern plates and replacing worn pins.

(b) Fins

A fin is a unwanted thin projection of metal found along parting line of casting. Fin may occur if the mould parts i.e., the drag and the cope are not clamped tightly and some gap is left between the mould parts.

Fin may be eliminated by placing sufficient weight on the top part i.e., cope of the mould so that two parts tightly fit together.

(c) Crush

Crush is an irregular shape depression on the casting surface. Crush occurs by the displacement of the sand at mould joints or core prints while closing the mould. Crush may also occur when large cores are pressed in too small core prints.

Defects due to Molten Metal

The defects occuring due to molten metal are pin hole porosity, mis runs and cold shuts, slag inclusions etc.

(a) Pin Hole Porosity

Pin hole porosity is due to the presence of hydrogen in the molten metal. The molten metal picks up hydrogen either in the melting furnace or by the dissociation of moisture in the mould cavity. During solidification, the molten metal looses temperature thereby the solubility of gases decreases and hence the dissolved gases will escape. The hydrogen while escaping the solidifying metal would cause very small diameter and long holes indicating the path of escape. These series of pin holes are called as pin hole porosity.

The pin hole porosity can be minimised by using good melting and fluxing practices, controlling the excess moisture in the moulding sand, increasing the permeability of the moulding sand and pouring the molten metal at a low pouring temperatures (gas pick up will be less if pouring temperatures are low).

(b) Mis Runs and Cold Shuts

Mis run occurs due to the failure of the molten metal to fill the mould cavity completely and thus leaving unfilled cavities. Cold shut occurs when two metal streams while meeting in the mould cavity, they do not fuse together properly thereby causing a discontinuity or weak spot in the casting.

Both misruns and cold shuts occur due to the presence of too thin sections. low fluidity of the molten metal and improper gating system. These defects may be eliminated by proper casting design and increasing the pouring temperature (gives high fluidity to the molten metal).

(c) Slag Inclusions

Any slag entering into the mould cavity along with the molten metal weakens the casting and also deteriorates the surface of the casting. Slag inclusions occur due to improper gating system and poor fluxing of molten metal. By rectifying these two factors, slag inclusions may be eliminated.

Defects Occuring During Solidification

Shrinkage cavities are formed by the liquid shrinkage occuring during solidification of the casting. Shrinkage cavities occur in the casting due to improper gating system, inadequate risers and poor casting design. This defect may be eliminated by achieving the directional solidification.

Metallurgical Defect

The main metallurgical defect is hot tears. Hot tears are cracks on the casting surface that develop when the normal contraction of the solidifying metal is obstructed. Hot tearing takes place between the temperature at which metal becomes coherent (i.e., the solidifying metal acquires some strength. Usually this happens when the metal is 50 to 90 per cent solid) and the end of freezing at the solidus. These cracks are called as hot tears because they occur at or just above the solidus.

Hot tears occur due to poor casting design, poor collapsibility of mould or core and hard ramming of the mould.

5.8.2 Inspection

Inspection determines the location and magnitude of various defects in the casting. Inspection methods are broadly classified as :

   (a) Destructive testing

   (b) Non destructive testing

Destructive testing involves picking up a specimen casting out of the given lot of castings and cutting it into two or more parts and examining for internal defects. The disadvantage of this technique is that several defects may go undetected unless the casting is sectioned at the place of defect and also the specimen casting may not be a correct representative and hence many defective castings may go undected.

Non destructive testing involves the use of physical principles to detect defective castings without destructing the castings. The various non destructive testing methods are as follows :

   (a) Visual examination

   (b) Penetrant test

   (c) Sound or percussion test

   (d) Radiography

   (e) Eddy current test

   (f) Magnetic particle inspection

   (g) Ultrasonic test

   (h) Pressure test

   (i) Impact test

Visual examination

To determine the surface defects, all the castings are subjected to visual examination. In the visual examination, the castings are illuminated with light and then examined with naked eye or by using photo cells and some times a low power microscope may be used as an aid to the eye. The main disadvantage of this test is that internal defects of the castings cannot be detected.

Penetrant test

Penetrant test helps in detecting small cracks which can not be noticed during visual examination.

In penetrant test a liquid penetrant usually kerosene oil is applied on the casting surface. This oil enters the cracks on the surface due to capillary action. The oil is thoroughly wiped and cleaned from the casting surface. Then the surface is white washed with $CaCo_3$ powder. After some time, the oil that has entered the defects will seeps out reducing the whiteness of the white wash of the surface. This reduction in whiteness will locate the nature and magnitude of the defects present in castings. The main limitation of this method is that only defects which are open to the surface are revealed.

Sound or percussion test

This is an ancient method. In this method, the castings are suspended with the help of chains and given blows with a hammer. By striking with a hammer, the defect free casting emits a clear ringing sound whereas the defective casting gives a dull sound. This method is applicable for simple castings since intricate shapes modify the sounds and confuse the inspector. The main disadvantage of this method is that it is difficult to determine the location and magnitude of the defect in casting.

Radiography

Radiography is used as a non destructive test to detect internal defects of the castings. Electro magnetic waves such as X-rays or gamma rays which have low wave lengths (10^6 - 10^{10}cm) are used in radiography as a means of inspection. When these electromagnetic waves pass through the casting containing defects are absorbed to a lesser extent than the rays passing through the adjacent defect free portions of the casting. After passing through the casting, the rays are allowed to fall on a film held against the opposite surface. Defects in the form of cracks or voids are recorded as blackened areas on the film since X-rays or gamma rays are absorbed to a lesser extent in these areas i.e., less dense regions.

The recorded film is called exograph if X-rays are used and a gamma graph if gamma rays are used.

Defects readily detected by radiography are sand inclusions, cracks, internal and external hot tears, unfused chills, shrinkage, gas or pin hole porosity - in short any defect exceeding about 2 to 3 per cent of the casting thickness.

Eddy current test

In this test, Eddy currents are induced by bringing a coil carrying alternating current near the casting. The path of the eddy currents is distorted by the presence of a defect and causes variations in the impedance of the coil. The change in the impedance, which gives an information about the magnitude of the defect, is analysed with the help of oscilloscope or recorder. Tllis technique detects cracks, voids and inclusions in castings.

Magnetic particle inspection

This test is used for iron and steel castings which are magnetic in nature. The principle involved in this test is that if a casting is magnetized, defects such as blow holes, cracks and inclusions produce a distortion in the induced magnetic field. By the application of a fine powder of magnetic material, this distortion of magnetic field can be detected. Hence this test consists of magnetizing the casting and then sprinkling the fine powder of magnetic material. This powder tends to be held and bridge over defects, thus forming a visible indication of the location and magnitude of the defect.

Ultrasonic test

Sound waves which have a frequency more than 20000 cps are called ultra sonic waves. In this test, a beam of ultra sonic waves are passed through one end of the surface of a casting, the waves travel through the casting to the opposite surface and are reflected back to the original point. Any defect in the path of the waves scatter the waves and are reflected back sooner from the defect than the waves passing through the defect free portion of the casting. For detecting the lengths of time, an oscillograph may be used which gives an indication of the location and the magnitude of the defect.

This test is used commonly for detecting internal voids in both ferrous and non ferrous castings

Pressure test

This test is used to detect the location of leaks and to test the overall strength of certain castings such as cylinders, valves, pipes and fittings, which are required to hold or carry fluids in service under different pressures. In this test, the casting is immersed in a soapy water containing in a tank and then the air pressure is applied. If there is a defect, air bubbles are formed at the defect place exactly.

Impact test

In this test a proper sized hammer is used to strike or fall on the region of the casting where some defect is suspected. A defective casting containing harmful defects is expected to break where as a defect free casting will not break. This test is not very reliable because sometimes even the defect free castings may break. Hence this method is rarely used.

5.9 MODEL QUESTIONS

1. List out various pattern materials and discuss their advantages and disadvantages.

2. Explain in detail various allowances given to a pattern.

3. What are various considerations to be taken while designing a pattern ?

4. With neat sketches, explain different types of patterns.

5. What are the constituents of a moulding sand and explain the function of binder in moulding sand.

6. List out the basic requirement of a moulding sand and explain them in detail.

7. List out various tests performed on a moulding sand and explain them in detail.

8. Explain the following:
    (i)   Pit moulding
    (ii) Machine moulding

9. Differentiate between jolting and squeezing.

10. Write short notes on the following:
    (i)   Green sand mould
    (ii) Dry sand mould
    (iii) Skin dried mould
    (iv) Loam mould

11. Explain carbon dioxide moulding method.

12. Briefly discuss about plaster moulding and metallic moulding.

13. With neat sketch explain the steps involved in a shell moulding process.

14. Explain the following :
    (i)   Ceramic moulding
    (ii) Centrifugal casting
    (iii) Slush casting

15. Differentiate between moulding sand and core sand.

16. What are the requirements of a core sand?

17. Explain the steps involved in the production of cores.

18. With neat sketch, explain the principle involved in core blowing machine and core shooting machine.

19. Explain various types of cores with neat sketches.

20. Write notes on the following :
    (i)   Vertical core
    (ii) Balanced core
    (iii) Hanging core
    (iv) Wing core

21. Explain the function of core prints and chaplets.

22. With neat sketch explain the operation of a cupola furnace.

23. Explain the two types of crucible furnaces with diagrams.

24. Show the components of a gating system with a diagram and explain them in detail.

25. What are the basic requirements of a sprue?

26. Explain the following:
    (i)   Top gates
    (ii) Bottom gates

27. Write notes on the following :
    (i)   Skimbob
    (ii) Skimming gate
    (iii) Whirl gate

28. Define gating ratio. Explain pressurized and unpressurized gating system.

29. What is the function of riser? Write note on the shape and size of risers.

30. Explain the following:
    (i)   Chvorinov's rule
    (ii) Caine's Method

31. Using caine's method, how do you find whether the given riser is a suitable one or not?

32. Explain the various methods which will improve the efficiency of a riser.

33. Explain the following:
    (i)   Padding
    (ii) Exothermic riser compounds

34. What is the function of a chill? Explain with neat sketches internal chills and external chill.

35. Explain various cleaning methods employed for castings.

36. Write short notes on the following :
    (i)   Plasma arc cutting
    (ii) Sand blasting
    (iii) Shot blasting
    (iv) Hydro blasting

37. Explain in detail any ten casting defects.

38. Explain the following:
    - (i) Blow holes
    - (ii) Scab
    - (iii) Rat tails
    - (iv) Mould shift
    - (v) Mis run
    - (vi) Hot tears

39. Differentiate between destructive and non-destructive testing of castings.

40. List out various non-destructive testing methods and explain radiography and ultrasonic test.

6 WELDING

Welding is a process of joining different materials mostly metals to produce a single piece of metal. It is a fabrication process in which portions of metal plate, castings, rolled sections and other metal pieces are joined to produce required shapes. Welding is defined as a process by which two similar or dissimilar metals may be joined by heating them to a suitable temperature with or without the application of pressure and with or without the use of filler metal.

6.1 TYPES OF WELDED JOINTS

The type of joint is determined by the relative positions of the two pieces which are being joined together. The most common types of joints are (a) Butt joint, (b) Lap joint, (c) T- joint, (d) Corner joint and (e) Edge joint.

(a) Butt Joint

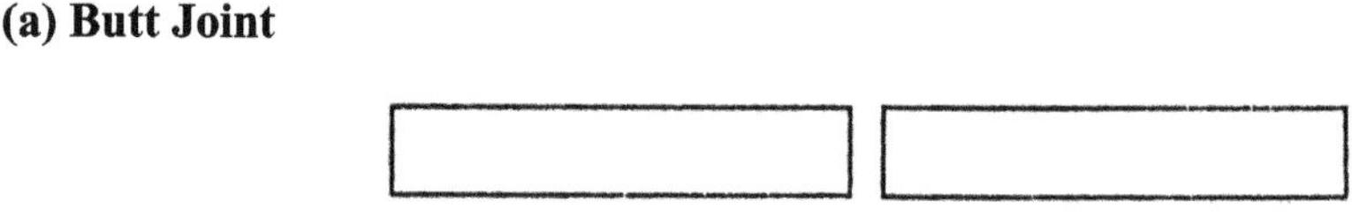

Fig. 6.1 *Butt Joint*

Fig. 6.1 shows the butt joint which is used to join the each of edges of two plates or surfaces located approximately in the same plane with each other.

(b) Lap Joint

Fig. 6.2 shows the lap joint which is used to join two over lapping plates so that the edge of each plate is welded to the surface of the other. Single lap or double lap joints are commonly used.

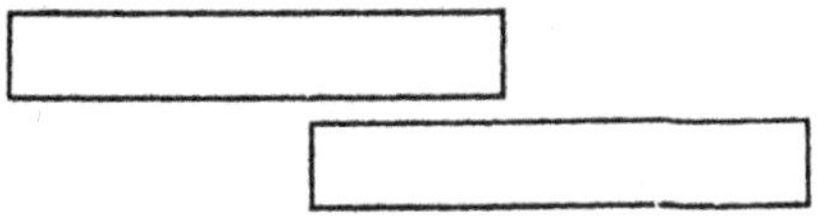

(c) T-Joint

Fig. 6.3 shows the T-Joint which is used to weld two plates or sections whose surfaces are at approximately right angles to each other.

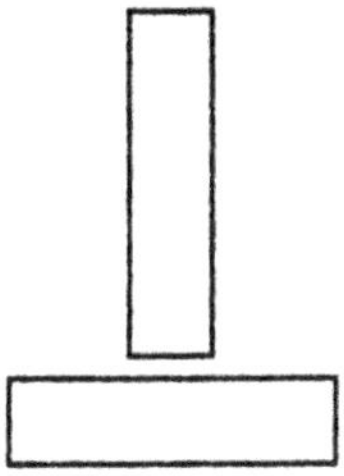

(d) Corner Joint

Fig. 6.4 shows the corner joint which is used to join the edges of two sheets or plates whose surfaces are at approximately 90° to each other.

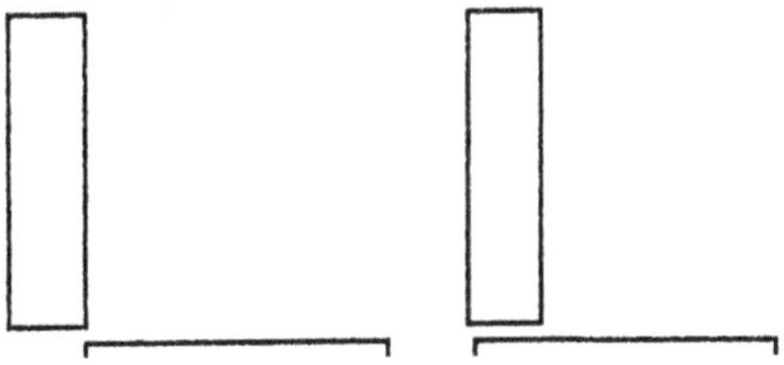

(e) Edge Joint

Fig 6.2 Lap joint.

Fig 6.3 T-Joint.

Fig 6.4 Corner Joint.

Fig. 6.5 shows the edge joint which is used to join two parallel plates. This joint is often used in sheet metal work.

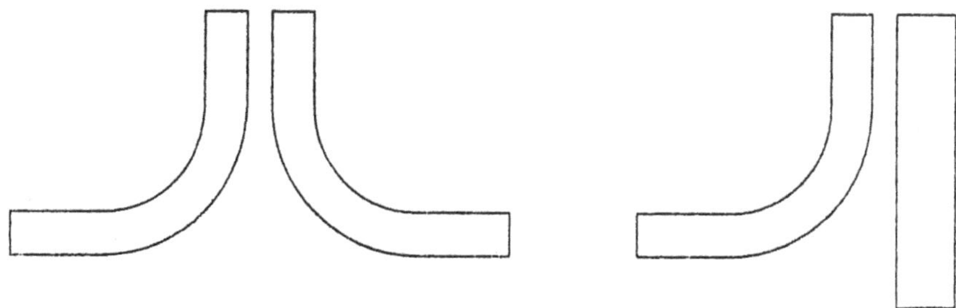

Fig 6.5 Edge Joint

6.2 CLASSIFICATION OF WELDING PROCESSES

Various welding processes are classified as follows :

    (i) Gas welding

        (a) Oxyacetylene gas welding

        (b) Oxyhydrogen gas welding

(ii)   Arc Welding

    (a)   Metal arc welding

    (b)   Carbon arc welding

    (c)   TIG or GTAW welding

    (d)   MIG or GMAW welding

    (e)   Submerged arc welding

(iii)   Resistance welding

    (a)   Spot welding

    (b)   Seam welding

    (c)   Projection welding

    (d)   Flash butt welding

(iv)   Solid state welding

    (a)   Cold pressure welding

    (b)   Friction welding

    (c)   Ultrasonic welding

    (d)   Forge welding

(v)   Thermo-Chemical welding

    (a)   Thermit welding

    (b)   Atomic hydrogen welding

(vi)   Special welding processes

    (a)   Electron beam welding

    (b)   Electro slag welding

    (c)   Stud welding

    (d)   Explosive welding

6.2.1 Gas welding

Gas welding is a fusion welding process which joins the metal using the heat of combustion of an oxygen/air and fuel gas (such as acetylene, hydrogen, propane) mixture. The intense heat of the flame thus produced melts and fuses together the edges of the parts to be welded. Usually filler metal is added to the weld to fill the space between the pieces being joined. Normally filler metal is added by inserting it into the molten puddle of the base metals and the puddle then solidifies making the weld bead as shown in Fig. 6.6.

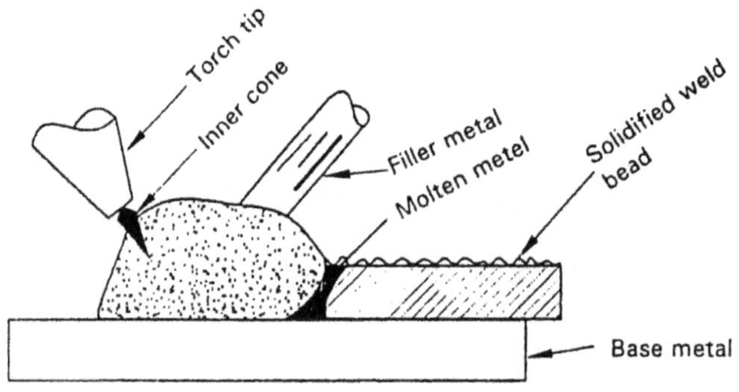

Fig 6.6 Gas Welding

6.2.1.1 *Oxy-acetylene Welding*

When acetylene (C_2H_2), which is produced by the chemical reaction between water and calcium carbide (CaC_2), is mixed with oxygen in correct propositions in the welding torch and ignited, the acetylene burns to produce a flame which is sufficiently hot to melt and join the base metal.This flame is known as oxyacetylene flame. The oxyacetylene welding system is shown in Fig. 6.7.

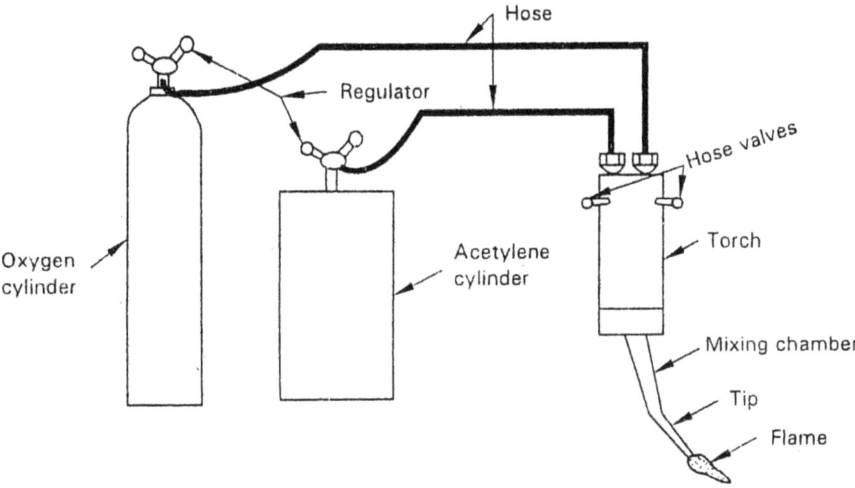

Fig. 6.7 Oxyacetylene Welding System.

The combustion of acetylene with pure oxygen at the tip of a gas welding torch occurs in two stages.

In the first stage, the following reaction takes place within the small, inner, brilliant white cone flame, close to the tip opening :

$$C_2H_2 + O_2 \rightarrow 2CO + H_2 + Heat$$

The heat produced by the above reaction is the most concentrated one giving highest temperature for welding.

In the second stage, the products of the first stage reaction ie., CO and H_2, react further by combining with the oxygen from surrounding atmosphere and gives :

$$4CO + 2H_2 + 3O_2 \rightarrow 4CO_2 + 2H_2O + Heat$$

This reaction takes place in the larger blue flame, which surrounds the small whitish cone flame. The heat produced by the second stage reaction is not as concentrated as the first stage reaction. Hence only a preheating effect is contributed by the larger blue flame and provides protective reducing atmosphere over the molten metal.

Ignition of a flame : Using the control valve on the welding torch, the acetylene is first turned on and then it is ignited with a friction spark lighter. The acetylene control valve is then adjusted until the flame ceases to smoke.

The oxygen control valve of the welding torch is then opened and adjust the proportions of acetylene and oxygen to the required level. This adjustment of acetylene and oxygen results in three distinct types of flames (Fig. 6.8) which are discussed below :

Types of flames : The three types of flames obtainable in oxyacetylene welding are :

(a)   Oxidizing flame (excess of oxygen)

(b)   Neutral flame (acetylene and oxygen in equal proportions)

(c)   Carburising or reducing flame (excess of acetylene)

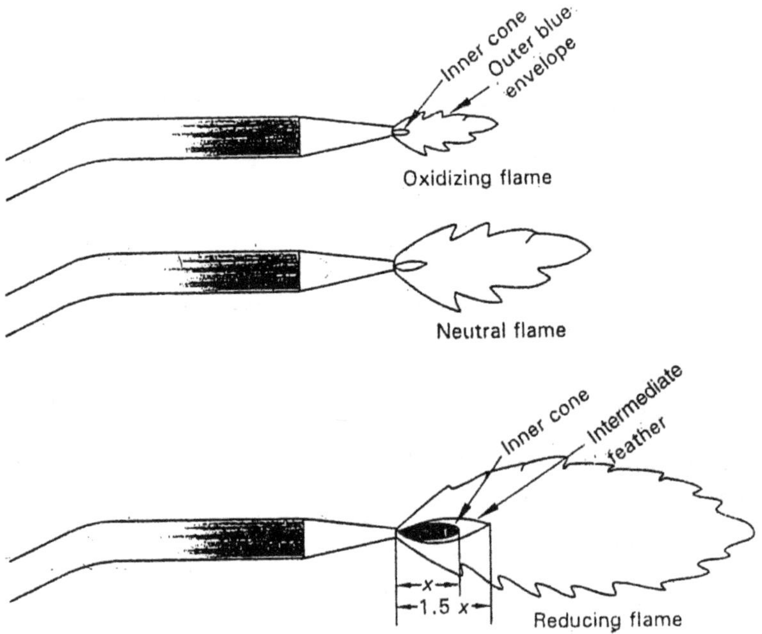

Fig. 6.8 Three types of flames.

Flame is the most important tool in oxyacetylene welding and all the remaining welding equipment simply serves to maintain and control the flame. To produce satisfactory welds, the correct type of flame i.e., proper size and shape should be chosen.

(a) Oxidising flame : Oxidising flame is obtained when there is an excess of oxygen than that required for the chemical reaction. The temperature of the oxidising flame is of the order of about 6300 °F (3482 °C). This is the hottest flame that can be produced by any oxygen-fuel mixture.

The oxidising flame consists of small inner cone which is shorter, pointed and much bluer in colour. The outer blue envelope is also much shorter than the other two flames. The oxidising flame injects oxygen into the molten metal of the weld puddle causing the metal to oxidize or burn quickly. Also the excess oxygen causes the weld bead to have a dirty appearance. Hence oxidising flame can not be used in most applications of oxyacetylene welding. But a slightly oxidising flame is advantageous in welding copper base metals, zinc base metals and few ferrous metals such as manganese steels and cast irons.

(b) Neutral flame : Neutral flame is obtained when approximately equal volumes of oxygen and acetylene are supplied. The temperature of the neutral flame is of the order of about 5900 °F (3260 °C)

The neutral flame consists of inner cone which is slightly larger than in case of oxidising flame. The outer blue envelope of neutral flame is also larger than in case of oxidising flame. Neutral flame does not cause any chemical change by not oxidising or carburising the metal and produce weld bead with clean appearance.

The neutral flame is commonly used for welding most of the metals such as mild steels, stainless steels, cast irons, copper, aluminium etc.

(c) Reducing flame : Reducing flame is obtained when there is an excess of acetylene than that required for chemical reaction. The temperature of the reducing flame is of the order of about 5500 °F (3038 °C)

The reducing flame is distinct in appearance than the other two flames by consisting of an inner cone and an intermediate feather. The outer envelope is also larger than the other two flames. The burning temperature of reducing flame is lower since it does not consume all the available carbon. The left over carbon is forced into the metal thereby causing the solidified weld bead to have pitted surface. However when welding high carbon steel, this excess carbon is an ideal condition.

WELDING TORCH

Fig. 6.9 shows all the features of an oxyacetylene welding torch. Welding torch mixes oxygen and acetylene in the required propositions and burns the mixture at the end of the tip. The mixer not only mixes oxygen and acetylene but also arrests any flash backs which occur if the flame travels back through the tip. The mixed gases coming out of the mixer are fed to the tip of the torch where the diameter of the orifice is chosen to match the volume of acetylene and oxygen mixture, so that a stable flame will be maintained. The tips are usually made with copper since copper possess high thermal conductivity. The central hole of the tip should be circular and smooth so that a streamline or laminar flow of gases can be maintained.

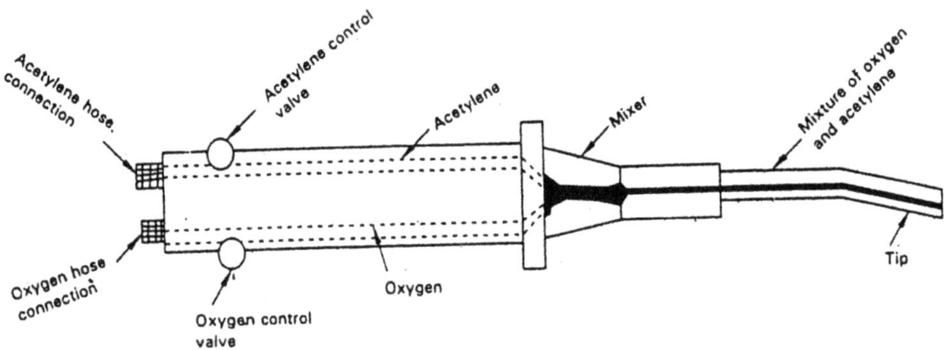

Fig. 6.9 An Oxyacetylene Welding Torch

Welding Techniques :

There are two usual techniques in gas welding depending upon the way in which the welding torch is used. These are :

(a) Left ward technique or Fore hand welding

(b) Right ward technique or Back hand welding

In the left ward technique, the operator can concentrate the heat from the torch ahead of the weld bead or in the weld puddle (Fig.6.10(a)). Left ward technique is usually used for welding relatively thin metals since the torch points in the same direction that the weld is being done so that the heat is not flowing into the metal as much as it could.

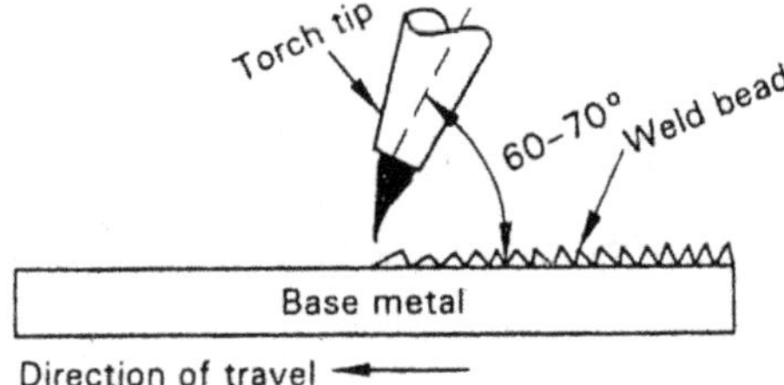

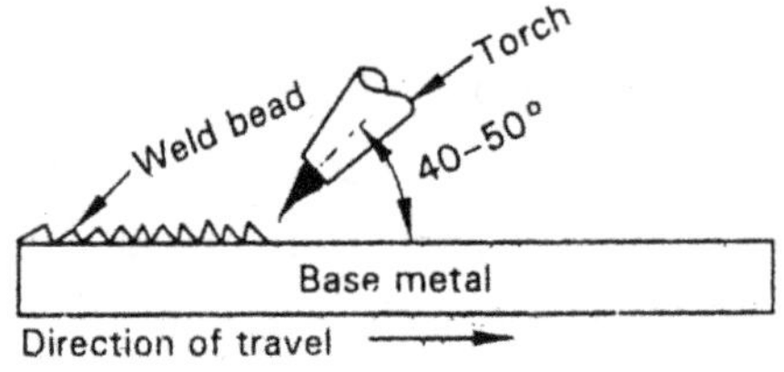

Fig. 6.10 Welding Techniques
(a) Left ward technique
(b) Right ward technique

In the right ward technique, the operator can concentrate the heat from the torch in the weld bead (Fig.6.10(b)). Right ward technique is usually used for welding thicker metals since the torch points in the direction opposite in which the weld is being done so that the heat is flowing into the metal as much as it could.

Gas Welding Rods

These are also called as filler rods. A filler rod is required for adding filler metal in gas welding. Usually correct welding rod should be used so that its composition and properties will match as closely as possible to those of base metal.

During welding, the welding rod melts and its metal is deposited in the puddle where it combines with the molten base metal to form strong weld. To suit the requirement of all types of metals and alloys, a variety of welding rods are available in its usual size varying from 1.6 mm to 10 mm diameter.

Gas Welding Fluxes

During welding, formation of oxides of certain metals produces poor quality, low strength welds or sometimes makes the welding impossible. Hence a flux which dissolves the oxide should be added to the welding area. Flux also acts as chemical cleaner, reducer and protector.

Fluxes for copper and copper base alloys consists of mixtures of sodium and potassium borates and boric acid. Fluxes used for ferrous metals are borax, sodium carbonate, sodium bicarbonate and sodium silicate.

Fluxes used for aluminium and magnesium consist of alkaline fluorides, chlorides and bisulphates.

After welding, all traces of flux should be removed from weld bead because all fluxes are chemically active and highly corrosive.

6.2.1.2 *Oxy hydrogen welding*

Low melting point metals such as aluminium, magnesium, lead etc., are welded using oxy hydrogen flame. Oxy hydrogen flame produces temperature of about 2500 °C maximum and hence this can not be used for welding steel. Oxy hydrogen flame is always reducing since there is no involvement of carbon during welding.

6.2.2 Arc Welding

Electric arc is used as the main source of heat in arc welding. Electric arc is produced when two conductors i.e. anode and cathode of an electric circuit are brought together and then separated slightly so that an air gap is established such that the current continues to flow through the gaseous medium. This arc produces temperatures of about 6000 –7000 °C.

Fig. 6.11 shows schematic diagram of an arc welding process. Either from an A.C. or a D.C source supply electric current. With A.C. due to the reversal of the current, the heat

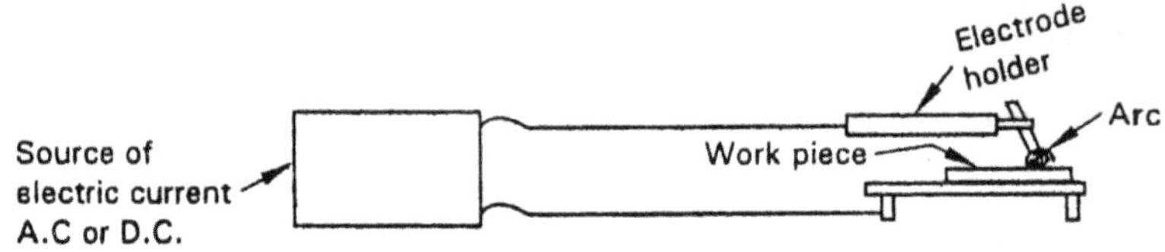

Fig. 6.11 *Schematic diagram of arc welding.*

generated at each pole is the same and hence changing over the connections to the electrode does not have any effect.

But the polarity of D.C. has great effect on electrode performance. With a D.C. source, if the work piece is connected to the positive terminal and the electrode holder is connected to the negative terminal of a welding machine, then the welding set up is said to have straight polarity.

And if the work piece is connected to the negative terminal and the electrode holder is connected to the positive terminal, then the welding set up is said to have reversed polarity.

Usually D.C. arc welding machines are D.C. generators which are driven by electric motors where as A.C. welding machines are transformers.

Both consumable and non consumable electrodes are used in arc welding. In the past bare electrodes were used but with bare electrodes it was found difficult to control the arc and to protect the molten metal from atmospheric oxygen and nitrogen which produces oxides and nitrides and makes the weld bead brittle and weak. Therefore modern electrodes are mostly coated. During welding the coatings on the electrode burns as the electrode melts and covers the molten metal from contamination.

Various arc welding processes are as follows :

   (a)   Metal arc welding

   (b)   Carbon arc welding

   (c)   TIG welding

   (d)   MIG welding

   (e)   Submerged arc welding

6.2.2.1 *Metal Arc Welding or SMAW*

This is conventional welding process where metal rod is used as electrode (consumable) and the work piece is used as another electrode. Fig. 6.12 shows the conventional metal arc welding process.

When the electric arc is struck between the electrodes, the base metal melts due to the heat of the arc. The consumable metal electrode will also melts and enters into the molten metal.

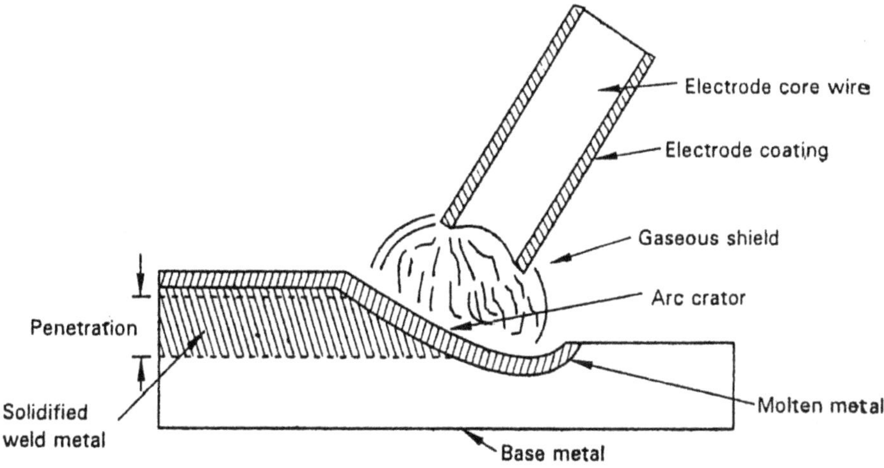

Fig. 6.12 Metal Arc welding process.

The depth to which the base metal is melted and deposited is called the penetration of the joint. The amount of penetration depends upon the cross sectional area of the weld bead, the current and the voltage. Penetration increases with increase in current.

A crater (small depression in the base metal) is formed due to the action of the electric arc. This arc crater should be avoided always otherwise it will become the weak point in the welded structure.

This process is also called as shielded metal arc welding (SMAW) because as the electrode coating burns it gives off an inert or protective gas which shields the arc and the molten metal from the atmospheric oxygen and nitrogen.

Both A.C. and D.C. current source may be employed in this method.

6.2.2.2 Carbon Arc welding

The fusion of metal is achieved in carbon arc welding by the heat of an electric arc which is struck between a carbon electrode and the work piece or between two carbon electrodes. Here a carbon or graphite rod is used as electrode instead of the consumable flux coated electrode. The carbon arc welding is usually used in the welding of aluminium, copper, nickel, monel and also some of the steel parts. The carbon arc welding utilises the power source from either D.C. or A.C.

The two types of carbon arc welding processes are as follows :

(a) Single Carbon Arc Welding

In single carbon arc welding, the arc is struck between the carbon or graphite electrode (negative) and the work piece (positive) (Fig. 6.13). The arc produced in this way heats the metal to the melting temperature. Always D.C. is preferred in this welding process than A.C.

Either graphite or carbon electrodes are used. Graphite electrodes are harder and brittle and they withstand higher current densities but it is difficult to control the arc. On the other hand carbon electrodes are softer and do not carry much current but control of the arc is easier. Carbon and graphite electrodes are considered to be non consumable electrodes but during the process of welding they deteriorate slowly due to vaporisation and oxidation.

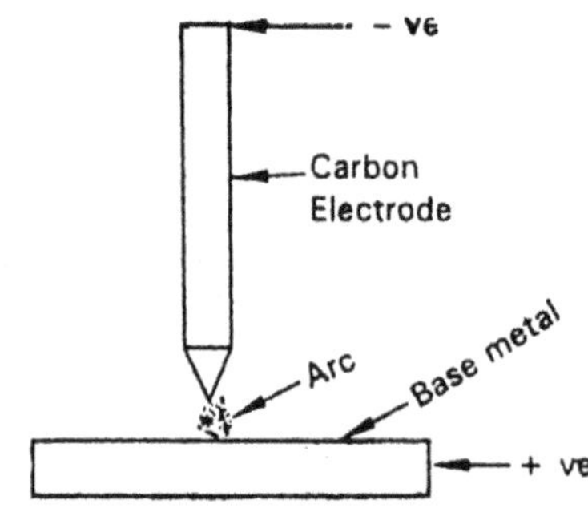

Fig. 6.13 *Single Carbon Arc Welding.*

Depending upon the type of joint and material to be welded, flux and filler metal may or may not be used.

(b) Twin Carbon Arc Welding

This process is similar to single carbon arc welding except that the arc is struck between two carbon electrodes instead of between carbon electrode and work piece (Fig.6.14).

In twin carbon arc welding A.C. source is preferred than D.C. because if D.C. supply is used the positive electrode will deteriorate and consume at a much faster rate as compared to the negative electrode since two thirds of heat is generated at positive electrode. This causes an unstable arc and requires frequent adjustment of electrodes. In case of A.C. source both the electrodes are affected equally due to alternate reversals of polarity and pose no problem.

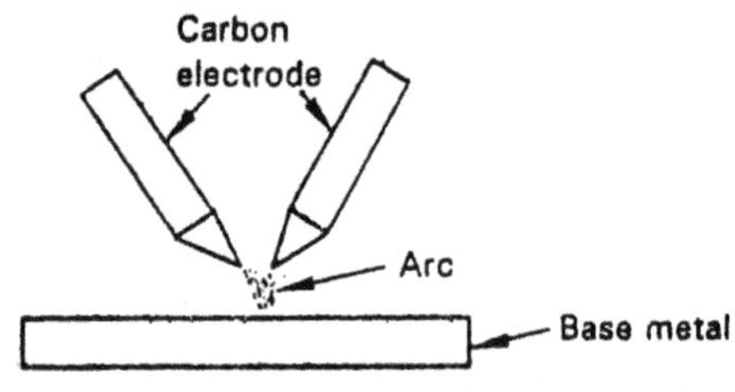

Fig. 6.14 *Twin Carbon Arc Welding.*

In this process, the work piece is not a part of the electrical circuit and the heat required to melt the metal is obtained by the electric arc which is struck between two carbon electrodes.

6.2.2.3 *Tungsten Inert Gas Welding (TIG) or GTAW*

Always the best weld is one that has the properties similar to those of the work piece. Therefore, during welding the molten puddle should be protected from atmospheric oxygen and nitrogen which makes weak weld beads. In TIG welding, an arc which is required to melt the metal at the joint, is struck between non consumable tungsten electrode and work piece in the presence of protective inert gas (Fig. 6.15). This inert gas protects the molten puddle from atmospheric oxygen and nitrogen. The inert gases that are used are argon, helium, and mixture of argon and helium.

The three basic power supplies which are used in TIG welding are direct current straight polarity (DCSP), direct current reversed polarity (DCRP) and alternating current high frequency (ACHF). ACHF is used only for starting purpose and switched off once arc is struck.

Pure tungsten or tungsten alloys are used as electrodes in TIG welding. The tungsten alloys are thoriated tungsten i.e., alloyed with thorium

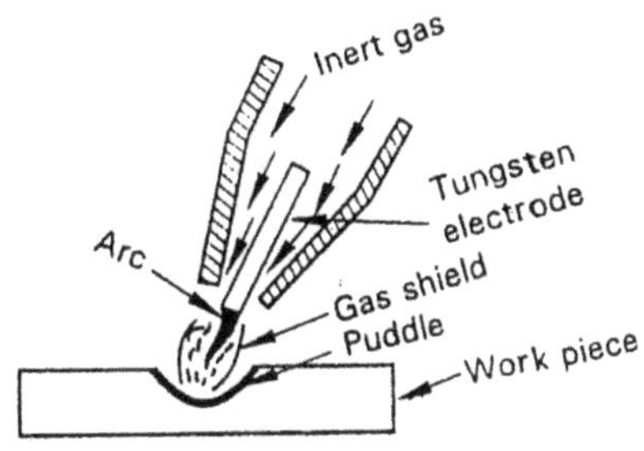

Fig. 6.15 *Tungsten Inert Gas Welding.*

and zirconiated tungsten i.e., alloyed with zirconium. Tungsten alloy electrodes possess high current carrying capacity, high resistance to contamination and produce stable arc.

TIG welding is used for welding of aluminium and its alloys, stainless steel, copper base alloys, magnesium alloys, carbon steel and low alloy steels.

6.2.2.4 *Metal Inert Gas Welding (MIG) or GMAW*

This is similar to TIG welding except the electric arc, required to melt the metal at the joint line, is struck between continuously fed consumable metal electrode and the work piece. The consumable electrode not only acts as the source of the electric arc but also supplies the filler material (Fig. 6.16). No flux is required in MIG welding because the molten puddle is shielded by the inert gas such as argon or helium or CO_2 gas. All ferrous and non ferrous metals can be

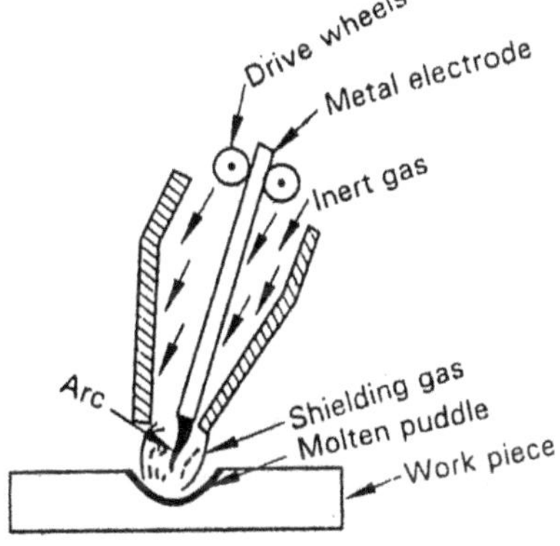

Fig. 6.16 *Metal Inert Gas Welding.*

welded using argon or helium as protective inert gas. Carbon dioxide gas is used only for ferrous metal because it is cheap, costing approximately one tenth as much as argon or helium.

6.2.2.5 Submerged Arc Welding

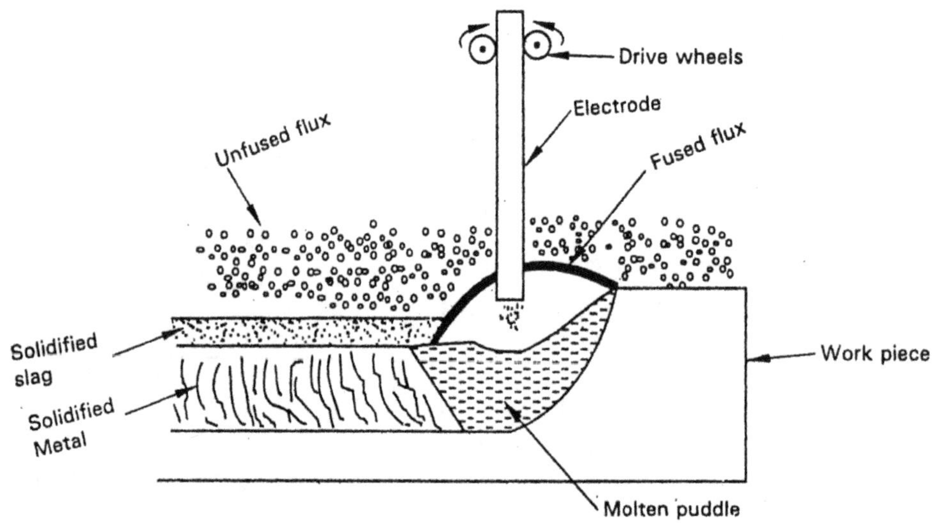

Fig. 6.17 Submerged Arc Welding.

This process is called submerged arc welding because the arc, the end of the electrode and the molten puddle are hidden or submerged in a finely divided granulated powder (flux). In this process, the arc required for melting the metal at the joint line is struck between a continuously fed bare wire electrode and the work piece. The flux powder is fed continuously from a hopper ahead of the arc zone. This flux protects the molten pool of metal from all effects of atmospheric gases thereby weld beads are exceptionally smooth. The flux adjacent to the arc melts and floats on the surface of the molten pool of metal and then solidifies to form a slag on top of the welded metal (Fig. 6.17). The remaining unmelted flux is recovered and reused. Normally, the flux contains silica, metal oxides and other compounds which acts as deoxidizers and cleaners.

The bare wire (or coated slightly with copper) electrode during welding melts and acts as filler. The electrode wire is continuously fed automatically from a reel and the electrode feed rate should match to the speed at which the electrode melts, thus keeping the arc length constant.

In submerged arc welding high currents are used. As a result the melting rate of the electrode and the speed of welding is very high. Either A.C. or D.C. power source can be used but above 1000 amperes A.C. power source is preferred to reduce arc blow.

Submerged arc welding is used for welding medium carbon steels, heat resistant steels. corrosion resistant steels, nickel, monel and many non ferrous metals.

This welding process produces high quality welds and also eliminates smoke and fumes since the arc column is submerged in flux, adds to its ease of operation and efficiency. This makes submerged arc welding to be used for a wide variety of applications in welding industry.

6.2.3 Resistance Welding

In resistance welding, the heat required to heat the metal to the plastic state over a limited area is obtained by resistance at interface of the work piece to the flow of electric current and the welding is completed by the application of pressure. No filler or fluxes are required during welding. Heat is generated by the contact resistance of the interface due to discontinuity of contact.

The amount of heat generated is related to the magnitude of electric current, the resistance to the electric current and the time of the current flow as expressed below :

$$H = I^2RT$$

where   H is the heat generated in joules

I is the current in amperes

R is the resistance in ohms, and

T is the time of current flow in seconds

It is very well established by Joule that a poor conductor heats up more than a good conductor for the same amount of current. Usually the interface between the two surfaces of the work piece offers the greatest resistance to the current flow due to poor conduction, compared to the rest of the circuit. Hence the interface area becomes the area of greatest heat.

A.C. power source is found to be most convenient for resistance welding since it is possible with A.C. to obtain any required combination of current and voltage. Usually current of the order of 3000 -100000 amperes is required to heat the metal pieces to the welding temperature and this can be obtained from a step down transformer.

It is established that the two important factors responsible for resistance welding are:

(a) The generation of heat at the weld joint by the resistance to the current flow.

(b) The application of pressure at the joint by electrodes.

The required pressure is exerted by the electrodes by employing springs, levers, cams or hydraulically or pneumatically controlled electrode arms.

The various resistance welding processes are :
  (i)   Spot welding
  (ii)   Seam welding
  (iii) Projection welding
  (iv)   Butt welding

6.2.3.1 *Spot Welding*

Spot welding is a resistance welding process which is employed to weld overlapping sheets at one or more spots by the heat produced due to the resistance to the flow of electric current through the work pieces which are held under pressure by two water cooled copper electrodes.

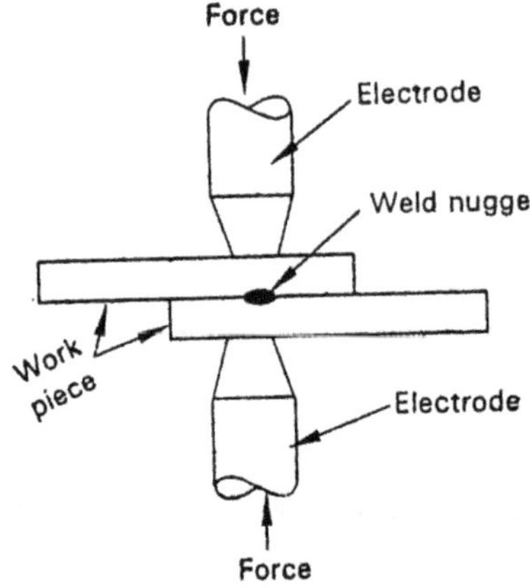

Fig. 6.18 *Spot Welding Process.*

The spot welding process is carried by first bringing the electrodes together against the overlapping work pieces and pressure is applied so that the two work pieces come in physical contact. Welding current is then switched ON for a fraction of second to a few seconds depending upon the nature of the material and its thickness (Fig. 6.18). As the current passes, a small area where the work pieces are in contact gets heated up to the welding temperature. Then the current is switched OFF and the pressure on the electrodes is maintained for a hold or forging time. After the hold time, the pressure is finally released, the electrodes are moved apart and the welded work piece is removed.

The water cooled electrodes must possess high thermal and electrical conductivity, considerable degree of hardness and the ability to maintain their shape at high temperatures. The electrodes for spot welding are made with hard copper alloys. The diameter of the fusion zone weld spot depends upon the diameter of the electrodes.

The surfaces of the work pieces must be clean before the application of pressure and the surfaces contact area must be uniform to have intimate contact under pressure.

6.2.3.2   *Seam Welding*

Seam welding is similar to spot welding except that in seam welding the joint is continuously welded instead of spot welds with the help of two copper alloy roller or wheel electrodes (Fig. 6.19). These rollers also apply the required pressure and the seam welds are produced by making high intensity current ON and OFF continuously. Gas filled electronic tubes such as ignitron and thyratron tubes are employed to switch the current ON and OFF continuously. Thus a continuous spot welds are produced by rotating the electrodes on the work piece which resembles to stitches. This process is called seam welding because it makes welds sufficiently close so that they overlap and make a leakproof seam.

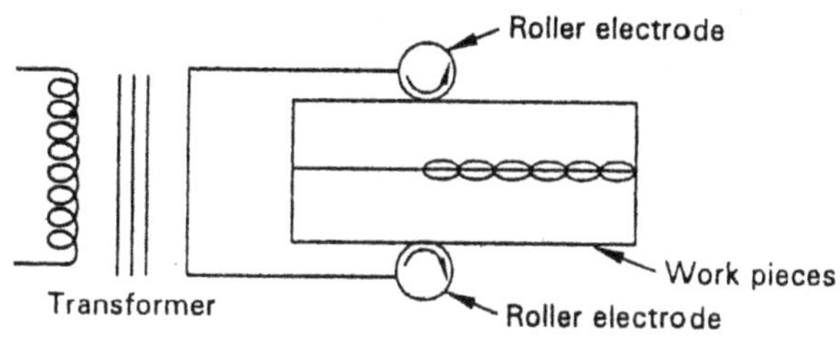

Fig. 6.19 Seam Welding.

6.2.3.3 *Projection Welding*

In both spot welding and seam welding current concentration is localized at the interface by shaping the electrode and restricting the area of the surface with which the electrodes are in contact. But in case of projection welding the weld areas are located by projections raised on the surface of one of the sheets. The work pieces touch only at the point of the projection when placed together (Fig. 6.20).

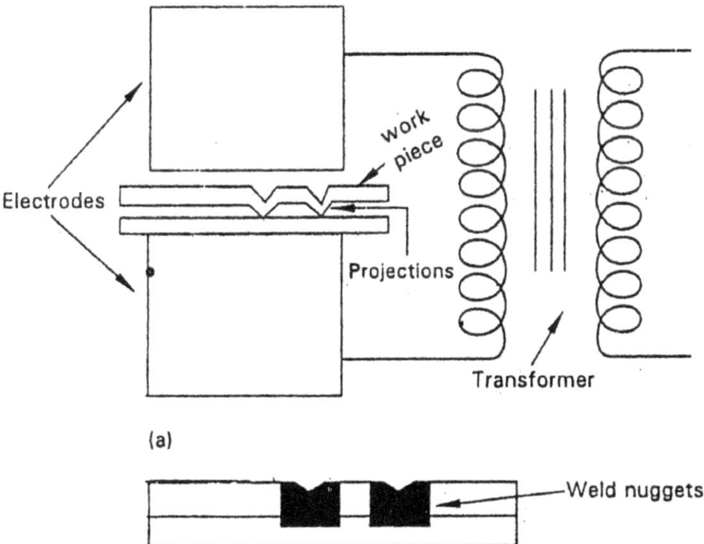

Fig. 6.20 (a) Projection welding process (b) Finished weld.

Projection welding is carried out by first bringing the work pieces together under electrode pressure. The current is then switched on and current flows across the interface which is concentrated around the projection. The metal in the projection area gets heated up to the plastic state and then the softened projection collapses under the pressure of the electrodes forming the required weld.

Flat copper alloy electrodes are used which are larger in diameter than the electrodes used in spot welding and hence the life of projection welding electrodes are much longer than the life of spot welding electrodes.

Projection welding requires less amount of current and pressure and hence it is adopted for many manufacturing processes.

In projection welding several welds can be made simultaneously but the ideal number of projections is three and maximum number of projections which can be welded satisfactorily are six.

6.2.3.4   *Butt Welding*

Welding two pieces of metals either on edge or on face is called butt welding. The important types of resistance butt welding are flash welding and upset welding.

(a) **Flash Butt Welding**

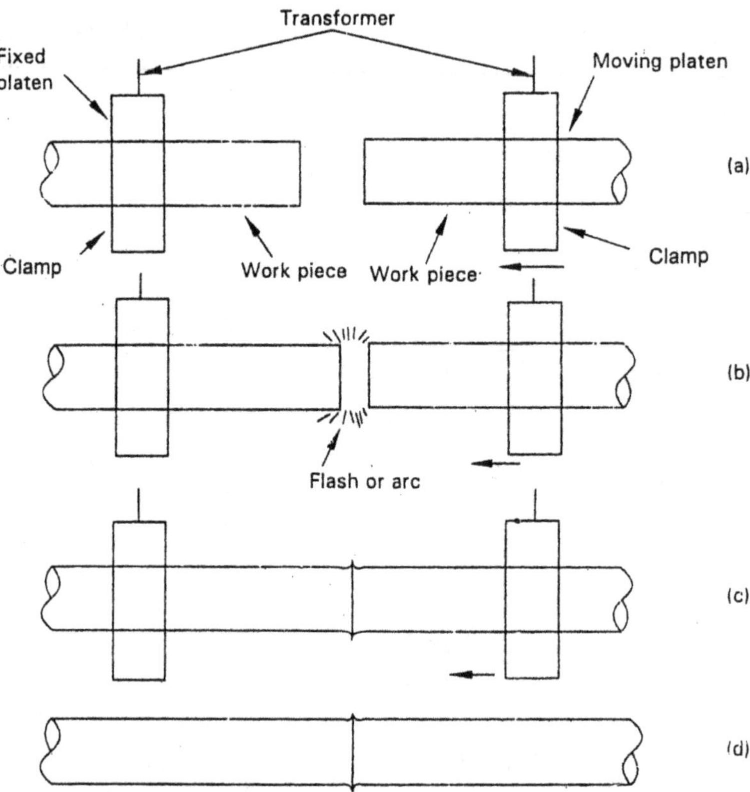

Fig. 6.21 *Flash Butt Welding (a) The work pieces are clamped in the circuit (b) Work pieces are brought closer with the help of moving platen and flash is produced (c) Moving platern further pressed to complete the weld (d) Finished weld.*

In flash welding an arc or flash is produced between the two work pieces to be welded.

Fig. 6.21 shows the sequence of operations in flash welding. First the work pieces are clamped in a fixed platen and a moving platen of the welding machine. Current is switched on and the two work pieces are brought closer. As the work pieces are moved closer, flashing or arcing will be produced. This arcing immediately raises the temperature of the work pieces to a welding temperature. Now the current is switched off and moving platen is further pressed to form a fusion weld expelling the molten metal and slag out of the weld joint. The expelled metal forms a fin or flash around the weld joint which is removed later by grinding or cutting.

Flash welding is used to weld rods, bars, tubing sheets etc. For most ferrous metals flash welding is best suited except cast iron.

(b) Upset Butt Welding

Upset butt welding is similar to flash butt welding and the difference is that in upset welding arc is not produced but the heat required for welding is generated due to the resistance to the flow of electric current. The equipment used is same as in flash welding i.e., one work piece is clamped to a fixed platen and the other to a moving platen. During welding, the two work pieces are brought together so that they contact each other (Fig.6.22). A heavy current is then switched on which causes an electrical resistance to be setup between the two work pieces thereby the interface heats up to the welding temperature. Then the two work pieces are pressed together firmly to form the weld. This pressing is called upsetting. Hence the high resistance and applied pressure are responsible for the formation of the weld joint in upset butt welding. Upset butt welding is used for commercial fabrication process.

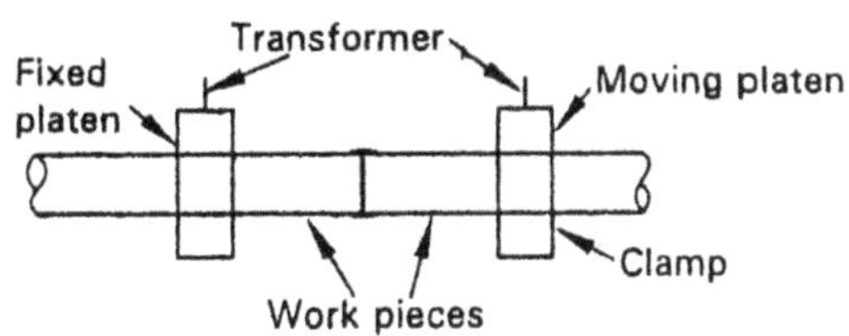

Fig. 6.22 Upset Butt Welding.

6.2.4 Solid State Welding

The various solid state welding processes are as follows :

  (i)   Cold pressure welding
  (ii)   Friction welding
  (iii) Ultrasonic welding
  (iv)   Forge welding

6.2.4.1   *Cold Pressure Welding*

In cold pressure welding, the weld joint is obtained by applying pressure at room temperature. Pressures ranging from 20000 to 200,000 psi are applied for welding. Under pressure the surfaces deform, breaking up the contaminants and brings areas of clean metal surface into intimate contact.

Cold pressure welding is used for some ductile non ferrous metals such as aluminium cables and connectors.

6.2.4.2   *Friction Welding*

In this process, the heat required for welding is generated from friction between the two surfaces. The work pieces are axially aligned so that one work piece can be rotated against a stationary part thereby producing the frictional heat by the speed of rotation and the axial pressure of the stationary work piece.

Fig. 6.23 shows the sequence of friction welding process. First the work piece is rotated and brought in contact with the stationary work piece under a light axial pressure. As the faces rub together friction produces the heat and soon the faces reaches the welding temperature. When sufficient heat is produced, the rotations is stopped and

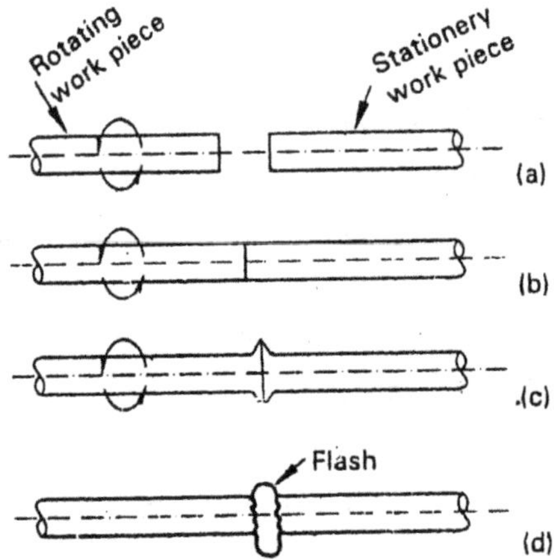

Fig. 6.23 *Friction welding sequence*
(a) Work piece is rotated (b) Rotated work piece moves and contacts stationary work piece (c) Slight pressure is applied (d) Finished weld.

the pressure may be increased with the help of stationary work piece. This results in weld joint having an excess metal flash which may be later removed by machining. Weld time is usually between 2 to 30 seconds depending on material to be welded. Applied pressure will be in the range of 5000 to 10000 psi and the speed of rotation varies between 1500 to 3000 r.p.m. Usually butt joints are produced by friction welding.

6.2.4.3   *Ultrasonic welding*

In ultrasonic welding, the weld joint is obtained by applying pressure and high frequency vibration motions. In this process, pressure is applied on two sides of the work while a hydraulic piston forces the welding pieces against a solenoid which induces ultrasonic vibrations into the work pieces. This results in the formation of cool, strong weld joint.

Ultrasonic welding is used for joining aluminium to steel, aluminium to tungsten, aluminium to molybdenum and nickel to brass. Ultrasonic welding finds unlimited applications in welding industry. To weld aluminium to stainless steel, ultrasonic welding is the only available method. This process is best suited for welding thin sheets and foils as thin as 0.0002 inches.

The time required to produce ultrasonic weld is same as the time required to strike an arc in arc welding.

6.2.4.4 *Forge Welding*

In forge welding, the metal edges to be joined are heated to the forging temperature and then forged together by applying pressure by hand or power hammering or pressing. This results in the formation of weld joint. The metals most commonly welded by forge welding are low carbon steels and wrought iron.

The most important application of forge welding is the manufacture of clad metal sheets and plates.

Forge welding is an oldest method and now a days this method is replaced by more faster, economical and latest welding processes.

6.2.5 Thermo-Chemical Welding

The various thermo chemical welding processes are :
  (i)   Thermit welding
  (ii) Atomic hydrogen welding

6.2.5.1 *Thermit Welding*

In thermit welding, a chemical reaction supplies the heat as well as filler material required for the weld joint.

Thermit welding is used for welding heavy sections. The ends of the heavy pieces to be welded are cut and aligned to provide a parallel sided gap. This gap is filled with wax, which will become the pattern for the weld. A suitable flask is placed around the wax pattern with the ends of work pieces extending to out side of the flask. The moulding sand is poured into the flask and then rammed and a gate, sprue, riser and a heating opening are formed in the sand. Heat is applied at the heating opening so that wax melts and burns out and also preheat the ends of the work pieces. After this the heat opening is plugged with a sand plug. Now the crucible above the flask is filled with thermit powder. For ferrous metals the thermit powder is a mixture of finely powdered aluminium and iron oxide. The thermit welding is based on the chemical reaction between finely powdered aluminium and iron oxide. The following exothermic reaction takes place once the thermit powder is ignited by igniting powder such as barium peroxide.

$$8Al + 3\ Fe_3O_4 \rightarrow 9\ Fe + 4\ Al_2O_3 + heat$$

This reaction gives molten iron, slag and heat. Usually the temperature obtained by this reaction will be above 3000 °C

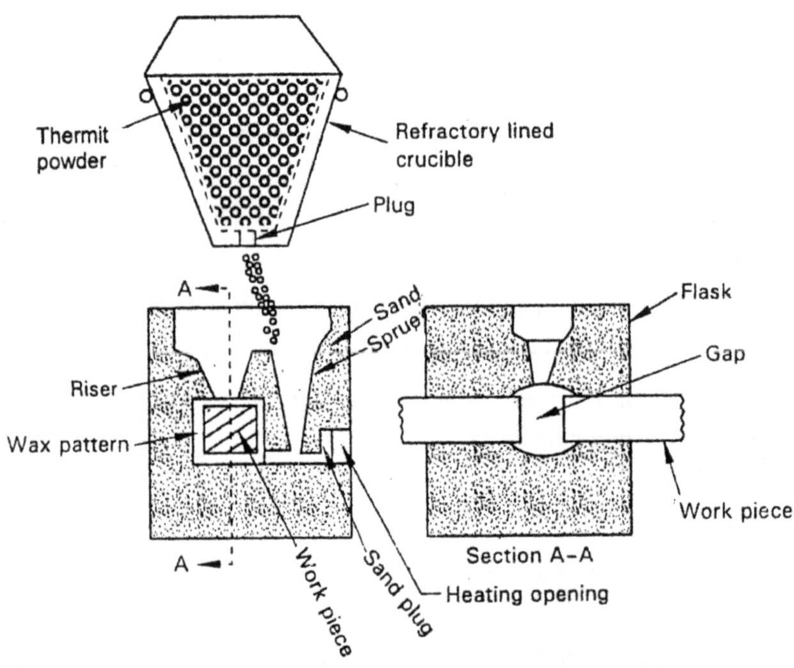

Fig. 6.24 *Sectional views of thermit welding process.*

Now, as shown in the Fig. 6.24, by removing the bottom plug of the crucible, the superheated metal produced by thermit reaction is allowed to flow into the mould between the ends of the work pieces to be welded. When solidified, the result will be the required weld joint.

Thermit welding is used usually for the repair of large broken parts.

6.2.5.2   *Atomic Hydrogen Welding*

In this process, the electric arc is struck between two non consumable tungsten electrodes above the work pieces to be welded in an hydrogen atmosphere. When the hydrogen gas passes near the arc, the hydrogen molecules dissociates into single atoms by absorbing the temperature of the arc. And the single atoms when approaches the relatively cool surface of the work pieces, they recombine releasing the large amount of heat which they have absorbed during dissociation. This recombination of hydrogen atoms produces the temperature required to join the work pieces. The hydrogen also serves as a shielding gas to the molten pool and protects from atmospheric contamination. When needed filler metal may be added during welding.

6.2.6 Special Welding Processes

The various special welding processes are :

    (i)   Electron beam welding

    (ii)   Electro slag welding

    (iii) Stud welding

    (iv) Explosive welding

6.2.6.1 *Electron Beam Welding*

Electron beam welding was developed during World War II in order to weld highly reactive and refractory metals such as titanium, molybdenum, tungsten etc.

In electron beam welding, the heat required for the weld joint is obtained by bombarding high velocity electron beam on to the work pieces to be joined. The operation is usually carried in vacuum otherwise the electron beam velocity reduces due to the electrons striking the small particles in the atmosphere.

In the electron beam process, the electrons are emitted from the cathode (heated tungsten filament) and these electrons travel through a series of focussing lenses and then the beam strikes the weld point (Fig. 6.25). When the beam strikes the welding point the kinetic energy of the high velocity electrons is converted into heat. This heat is sufficient to melt and fuse the metal. The greater the kinetic energy the greater the amount of heat released. Electron beam welding is highly successful in obtaining deep penetration with little distortion.

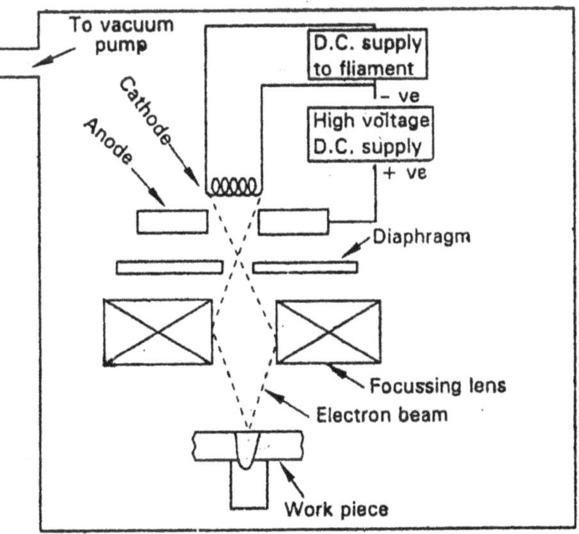

Fig. 6.25 Electron beam welding.

Because of the vacuum and since heat is obtained by electrons, electron beam welding completely eliminates contamination of both the weld zone and the weld bead. The equipment of electron beam welding is very expensive.

6.2.6.2 *Electro Slag Welding*

Electro slag welding is popular in the fabrication of heavy machine components and pressure containers with thick walls.

In this process the heat required for the welding is produced by the flow of current through, the consumable electrode wire and the molten slag, hence it is called as electro slag welding.

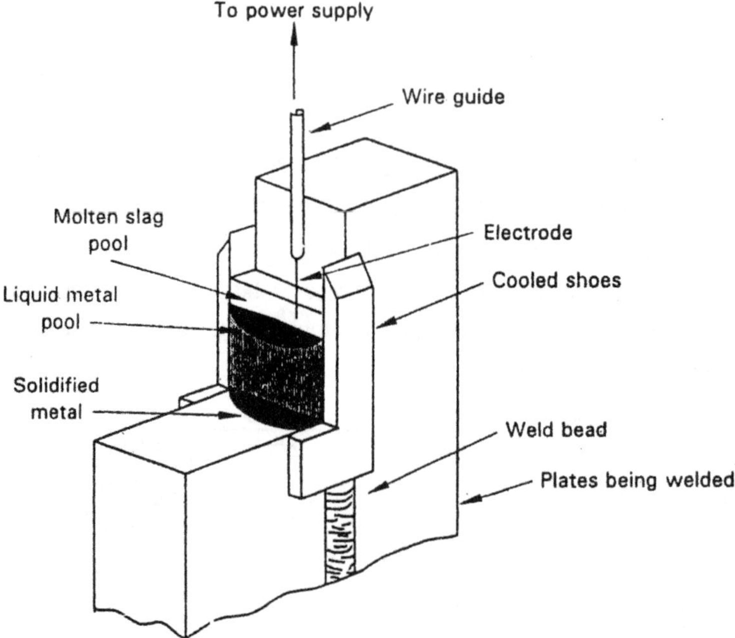

Fig. 6.26 Electro slag welding.

Electro slag welding is a vertical process where two copper shoes are provided on either side of the joint (Fig. 6.26). The space formed between the weld joint and the copper shoes contains molten slag pool into which the electrodes are immersed. The electro slag process is started by striking an electric arc beneath a layer of granular welding flux. The arc is stopped when sufficient flux is melted or molten slag is formed. Then the electric current passes from the electrode to the work piece through the molten slag. Now the heat will generate due to the resistance to the flow of current through the molten slag which is sufficient to melt the edges of the work piece and also to melt the welding electrode. The molten pool of metal then slowly solidifies resulting in a strong weld bead.

6.2.6.3 *Stud Welding*

In stud welding, the heat required for the weld joint is obtained by striking an electric arc between a metal stud, bolt, rivet or similar part and the base metal.

Stud to be welded is held in collet chuck and ferrule (non conductor of heat and electricity) which is made up of ceramic or porcelain material is placed around the welding end of the stud (Fig. 6.27(a)). This ferrule concentrates the heat of the arc in the weld zone, confine molten metal to the weld area, and protect the weld puddle from contamination from atmosphere.

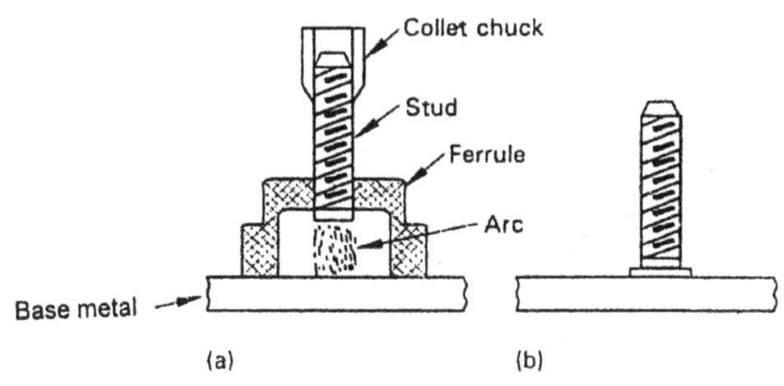

Fig. 6.27 *(a) Stud welding process (b) Finished weld.*

In stud welding, first an electric arc is struck between the stud and the base metal. The electric arc melts the base metal, the end of the stud and flux. Then the arc is stopped and the melted end of the stud is forced against the molten metal pool of the base metal. This causes a fusion weld to take place between the stud and the base metal as shown in Fig. 6.27 b.

Both ferrous metals such as carbon steel, stainless steel, low alloy steels etc., and non ferrous metals such as brass, bronze, aluminium can be stud welded.

6.2.6.4 *Explosive Welding*

Explosive welding is a solid state welding process, where the weld joint is made with high relative velocity at a high pressure which is obtained by detonation of an explosive such as RDX or TNT (Fig. 6.28). Here a large amount of plastic interaction occurs between the surfaces to be welded.

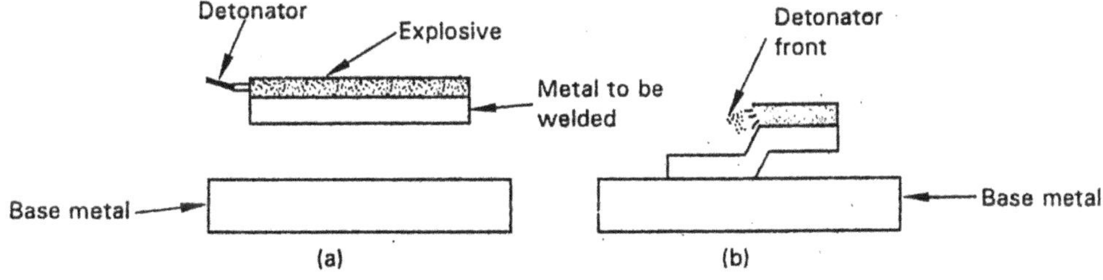

Fig. 6.28 *(a) Explosive welding set up (b) weld formed after detonation.*

In explosive welding, the detonation of an explosive produces both extremely high normal pressure and a slight, relatively shear or sliding pressure between the surfaces which causes the welding of the two surfaces. At the impact point the velocity may range from 500 to 1000 ft/sec with a pressure of about 100,000 to 1,000,000 psi. The explosive weld formed will be as strong as the base metal.

Explosive welding is used to join tungsten to steel, aluminium to stainless steel, aluminium to nickel steel, copper to stainless steel etc.

6.3 WELD SYMBOLS

Table 6.1 summarises various weld symbols

From of weld	Weld symbol	Form weld	Weld symbol
Fillet	◿	Spot weld	✳
Square Butt	⊓	Seam weld	✕✕✕✕
Single V Butt	▽	Projection weld	△
Double V Butt	✕	Stud weld	⊥
Single U Butt	∪	Flash weld	⋀
Double U Butt	8	Butt resistance or pressure (Upset)	ǀ
Single bevel Butt	⌐	Dead or edge weld	◠
Double bevel Butt	K		
Single J Butt	Ɗ		
Double J Butt	Ɓ		

Table 6.1 : Weld Symbols

6.4 WELDABILITY

Weldability is the capacity of a metal to be welded by forming a strong bond between the work pieces. Good weldability means the welded joint should perform satisfactorily during its service. This can be achieved if the weld joint is produced with out any defects.

Weldability of metals is affected by the following factors :

(a)   Effect of welding heat on base metal

    (b)   Effect of heat on alloy formation at the weld zone

    (c)   Eflect of base metal on weld deposit.

The weldability usually varies for different metals and alloys. Weldability of metals and alloys depends upon (i) melting point of metal and fusibility (ii) thermal conductivity of metals (iii) Thermal expansion (iv) surface conditions and strength (v) changes in microstructure.

All the factors that affect weldability adversely may be corrected by using adequate shielding atmosphere, proper filler metals, proper fluxing material, proper heat treatment of the metal before and after weld deposition and by using proper welding method. Hence by controlling various factors that effect weldability, almost all metals can be welded by one or other process of welding.

6.5 THERMAL CUTTING METHODS

Various metal cutting methods are as follows:

(A) GAS FLAME CUTTING METHODS :
    (i)   Oxy-acetylene cutting
    (ii)   Oxygen lance cutting

(B) ARC CUTTING :
    (i)   Carbon arc cutting
    (ii)   Metal arc cutting
    (iii) Arc-oxygen cutting

6.5.1 Gas Flame Cutting Methods

The gas flame cutting methods are oxyacetylene cutting and oxygen lance cutting. Gas flame cutting methods are based on the ability of certain metals mostly iron and steel, to burn in oxygen with the evolution of a large amount of heat. Usually iron and steel oxidizes when exposed to air but this process of oxidation can be enhanced by heating the steel to a red hot condition and then supplying oxygen to the hot spot. By doing this, the metal burns rapidly since the reaction between iron and oxygen is exothermic i.e., produces large amount of heat. This heat keeps the oxide which is formed due to oxidation in molten condition which flows or may be blown off thereby exposing more metal to the action of oxygen. This is the principle involved in gas flame cutting methods.

6.5.1.1   *Oxy-Acetylene Cutting*

Oxyacetylene cutting employs ordinary gas welding equipment except that the welding torch is replaced by cutting torch. The cutting torch slightly differs from welding torch. Cutting torch is fitted with two valves which regulate the supply of oxygen and acetylene which are mixed together for preheat flame and the method used for mixing is similar to welding torch. A third valve controls the stream of cutting oxygen. Fig. 6.29 shows a typical cutting torch tip which consists of four small openings surrounding a large opening at the centre. The large opening at the centre provides the cutting oxygen stream whereas the around it provide the mixture of oxy-acetylene gas.

In oxy-acetylene cutting, first the metal is preheated with a preheating flame which is produced by mixing oxygen and acetylene in correct proportions in the cutting torch. After a spot area along the line of cut is

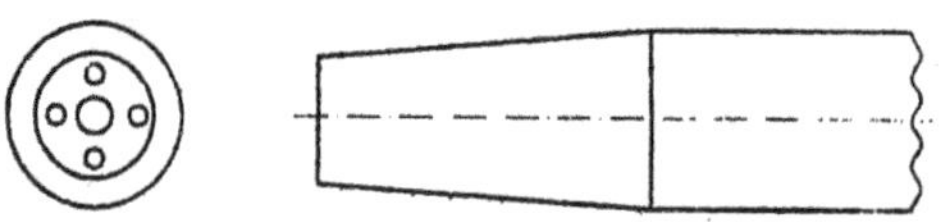

Fig. 6.29 Typical cutting torch tip.

heated to red hot, a jet of pure oxygen under pressure is directed to the heated area through the central opening of the tip of the cutting torch. This cutting oxygen jet penetrates through the steel and causes oxidation reaction between iron and oxygen resulting in the formation of iron oide. The reaction products of the oxidation reaction are blown off from the reaction zone by the pressure of the oxygen gas thereby exposing more metal to the action of oxygen thus progressively cutting a narrow slit or kerf in the metal along the line of cutting.

Oxy-acetylene cutting operation may be carried out either manually or in a mechanized way. In general, machine cutting in which the nozzle is traversed mechanically along the cutting line gives a higher quality of cut.

6.5.1.2   *Oxygen Lance Cutting*

Oxygen lance cutting is employed to cut heavy thick steel sections Which can not be cut by oxy-acetylene cutting

In oxygen lance cutting, the edge of the work piece to be cut is preheated with a preheating torch (Fig. 6.30). The oxygen is directed to the preheated area using oxygen lance (pipe) at a pressure of 40 to 50 psi. This causes rapid oxidation of the surface. The oxygen jet blows the reaction products thereby cutting the work piece along the required line of cutting.

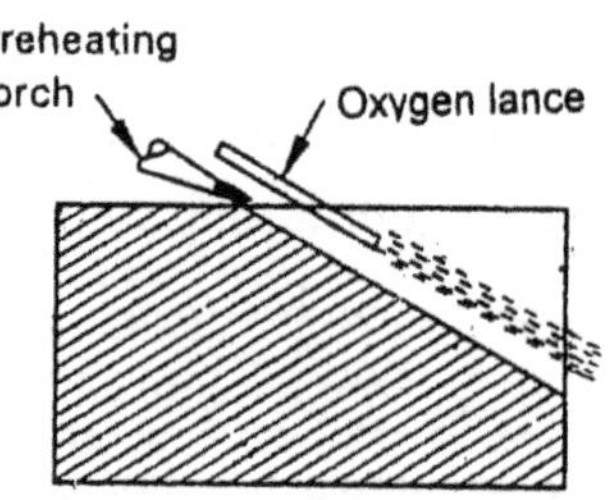

Fig. 6.30 Oxygen Lance Cutting.

6.5.2 Arc Cutting

In arc cutting, the cutting operation is performed by melting the metal with heat supplied by an electric arc which is struck between the electrode and the work piece, Some provision must be made to allow the molten metal to drop by gravity from the cutting area.

6.5.2.1 *Carbon Arc Cutting*

In carbon arc cutting, the equipment used is similar to that used in carbon arc welding. A graphite electrode is usually preferred and D.C. source with straight polarity is used. As usual the arc is struck between graphite electrode and the work piece. The electric arc melts the metal and the molten metal flows away from the cutting area felicitating further cutting of metal.

All metals which can be readily melted can be cut by this method. Advantage being low cost but the disadvantage is it produces roughest cut edges and hence this method is employed for cutting scrap metal.

6.5.2.2 *Metal Arc Cutting*

This is similar to carbon arc cutting except that a consumable metal electrode (special coated mild steel) is used. This coating on electrode prevents short circuiting at the sides of the electrode, also stabilizes the arc. As usual here also the arc is struck between the consumable metal electrode and the work piece. This electric arc melts the metal and the molten metal flows away from the cutting area exposing the metal to be cut.

Metal arc cutting is a costlier process than the carbon arc cutting since the electrodes are used up during cutting operation. Hence metal arc cutting is rarely used.

6.5.2.3 *Arc-oxygen Cutting*

This method is similar to oxy-acetylene cutting except that the preheating is done by an electric arc struck between consumable tubular electrode and the work piece. The coated tubular electrode possess a small hole at the centre through which oxygen is blown to the heated area or cutting area.

In this method, the cutting operation involves first striking an electric arc between tubular electrode and the work piece. This electric arc preheats the work piece at the cutting area. Then the oxygen is blown through the centre hole of the tubular electrode to the cutting area. This causes rapid oxidation of the metal. This oxidation reaction is an exothermic reaction i.e., it produces large amount of heat. This heat keeps the oxidation reaction products in the molten condition. During cutting these reaction products flows away from the cutting area exposing fresh metal to be cut and thus cutting progress towards the line of cutting.

The cut edges are smoother than carbon-arc cutting but not as smooth as those obtained in oxy-acetylene cutting.

6.6 BRAZE WELDING OR BRONZE WELDING

Braze welding, also called as bronze welding uses a bronze filler rod. The bronze filler rod usually consists of 60 per cent copper and 40 per cent zinc with suitable amounts of silicon and tin which act as deoxidisers.

In braze welding the edges are not melted but only heated by an oxy-acetylene flame to a temperature corresponding to the melting point of bronze filler rod. This temperature depends upon the composition of the bronze filler rod. After heating the cleaned edges to be joined to the braze welding temperature, flux is applied to the edges and then the filler rod is brought in contact with the heated edges. The filler rod then melts and adheres to the edges by virtue of surface tension or it tins the edges and forms a sound joint on solidification.

Braze welding is usually used on steel and cast iron and also joins copper, brass, bronze, nickel and nickel alloys.

The main advantage of braze welding is that less heat is required for making the joint and also dissimilar metals which can not be welded by conventional methods can be braze welded.

The disadvantage of braze welding is that braze welded joints does not give satisfactory service at temperatures above 250 °C and also for dynamic loads of about 1000 Kg/cm^2 or more.

Braze welding may be employed for fabrication work of metal furniture, automobiles, and other house hold appliances.

6.7 SOLDERING

The difference between welding and soldering is that in soldering there is no direct melting of the base metal but joining of the two metal pieces occur by filling the joint with filler metal called solder. The melting point of the solder is always much less than the base metal. The solder is usually an alloy of lead and tin, lead-tin-antimony, lead and silver. A suitable flux is always used in soldering to prevent oxidation of the joint. The usual fluxes are zinc chloride, organic fluxes which contain stearic acid, benzoic acid, lactic acid etc., and rosin flux which primarily consists of gum exuded from pine trees.

Many methods can be employed for heating the base metal and melting the solder and flux. But the widely used and the oldest method is by means of traditional tool called soldering iron with a copper tip which may be heated electrically or by gas flames.

In soldering process, first the surfaces of the parts to be joined are cleaned from oil, grease, scale and dirt thoroughly. Then the parts are arranged in the required alignment and position. The tip of the soldering iron is heated, dipped in flux and then rubbed on the solder to pick up a small quantity of solder with it and then deposited gently along the joint. The soldered joint is then cleaned to remove any undesirable flux residues.

Soldering usually used for making electrical connections, and repair work.

6.8 BRAZING

Brazing is similar to soldering. In brazing also the work pieces are not melted at the joint. The joining of two metal pieces takes place in brazing by heating to just above the melting temperature of filler metal which in this case known as spelter. The melting point of spelter is below the melting point of work pieces to be joined. The copper base alloys and silver base alloys are generally used as filler metal in brazing. A suitable flux such as borax and mixtures of borax and boric acid is used.

In brazing process, first the surfaces of the parts to be joined are cleaned from oil, grease, scale and dirt thoroughly. Then the parts are arranged in the required alignment and position. Flux is then applied to all surfaces where filler metal is to flow and the joint is heated to the brazing temperature (above the melting point of filler metal). Now, when the filler metal is applied to the joint, it flows throughout the joint by capillary action.The brazed joint is then cooled and cleaned to remove any flux residue.

Several methods are employed to apply heat for brazing. Either an oxy-acetylene torch, or an electric arc, or an induction heating generally be employed.

Brazing usually produce much stronger joints than soldering and is used to join a large variety of dissimilar metals.

6.9 MODEL QUESTIONS

1. Define welding process. What are different types of welded joints ?

2. With neat sketch explain oxyacetylene welding process.

3. Distinguish between oxidising flame, neutral flame and carburising flame.

4. Write a note on gas welding torch and gas welding techniques.

5. Differentiate between gas welding and arc welding?

6. Explain the following :
   (i)   Metal arc welding
   (ii) Carbon arc welding

6. Explain the following :

7. With neat sketch explain submerged arc welding process.

8. Draw a neat sketch and explain submerged arc welding process.

9. Explain the principle involved in resistance welding.

10. Differentiate between spot welding and seam welding.

11. With a neat sketch explain spot welding process.

12. Explain the following with neat sketches :
    (i)   Seam welding
    (ii) Projection welding
    (iii) Butt welding

13. Distinguish between flash butt welding and upset butt welding.

14. List out various solid state welding process and explain friction welding.

15. Explain the following :
    (i) Ultrasonic welding
    (ii) Forge welding

16. With a neat sketch explain the thermit welding process.

17. With a flow diagram explain electron beam welding process.

18. Explain the following :
    (i) Stud welding
    (ii) Explosive welding

19. What are the advantages of electro slag welding over conventional welding processes ?
    Explain electro slag welding process.

20. Write a note on the following :
    (i) Weld symbolds
    (ii) Weldability

21. Explain the following :
    (i) Oxyacetylene cutting
    (ii) Oxygen lance cutting

22. Differentiate between gas flame cutting and arc cutting.

23. Write a note on the following :
    (i) Carbon arc cutting
    (ii) Metal arc cutting

24. Write a note on braze welding

25. Distinguish between soldering and brazing.

7   PLASTICS

The term polymer is derived from two Greek words namely polus = many, mer = unit. A polymer is composed of a large number of repetitive units called monomers (mono = one). Normally a polymer consists of thousands of monomers joined chemically together to form a large molecule. A few terms are defined below for better understanding of polymers.

(a) Polymerization – It is the process of forming a polymer ie., joining the end to end of monomers.

(b) Degree of polymerization $= \dfrac{\text{Mol. wt. of a polymer}}{\text{Mol. wt of a single monomer}}$

(c) Linear polymer – A polymer which is obtained by simply adding the monomers together to form long chains is called Linear polymer

(d) Copolymer – A polymer which is obtained by adding different types of monomers is called copolymer

(e) Cross linked polymer – A polymer which is obtained by connecting the long chains through a covalent bond.

7.1 POLYMERIZATION

Polymerizallon is the process of forming a polymer. A polymerization reaction is the conversion of a particular compound to some large multiple of itself. Classically, polymerization reaction is divided into two main groups namely,

(a) Addilion polymerization .

(b) Condensation polymerization

7.1.1 Addition Polymerization

Addition polymerization is the process in which two or more chemically similar monomers (compounds) are polymerized to form long chain molecules. This occurs in unsaturated organic compounds i.e., molecules having double or triple bonds. Since these compounds have double or triple bonds, they are unstable as compared to saturated organic compounds. The unsaturated organic compounds, under suitable conditions react to form long chains i.e., addition polymerization occurs. During addition polymerization, their double covalent bond is broken and single bonds are formed in its place. An example of addition polymerization is shown in Fig. 7. 1

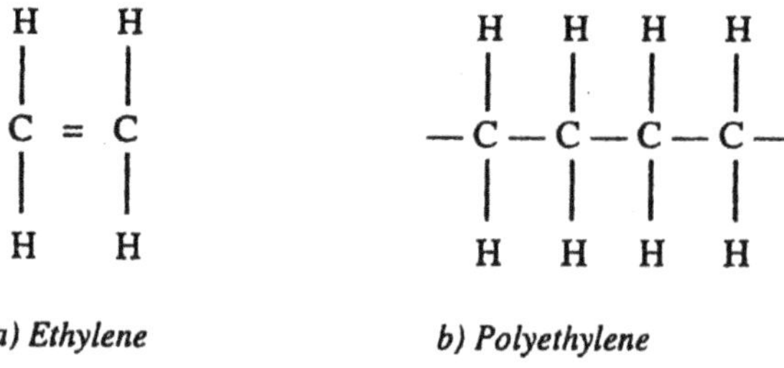

a) Ethylene   *b) Polyethylene*

Fig. 7.1 *Addition Polymerization.*

Fig. 7.1 shows an ethylene monomer. Now if one more ethylene monomer is added to this monomer, the double bonds are broken and single bonds are formed between carbon atoms. The new molecule formed as a result of addition polymerization is known as polyethylene (polythene). It is found that the addition polymerization process do not take place automatically when two or more monomers are added. But it requires some amount of energy in the form of heat, light, pressure or the presence of catalyst for its initiation. And another important condition is that the monomers must be available in the vicinity of the end of the chain or the unpolymerized monomers should continuously diffuse to the regions where they can get polymerize with other monomers.

Another example of addition polymerization is the formation of polyvinyl chloride (P.V.C) which is shown in Fig. 7.2.

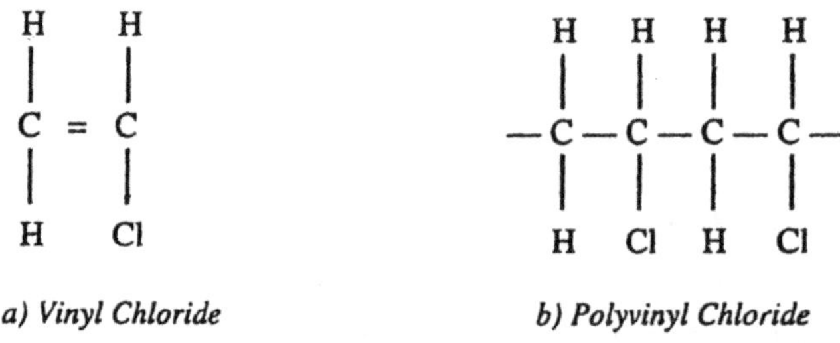

a) Vinyl Chloride   *b) Polyvinyl Chloride*

Fig. 7.2 *Addition Polymerization.*

It is found that the polymers which form the monomers are unstable because the carbon atoms at both ends of the long chain do not have four covalent bonds. So, to provide stability some terminal radicals such as H_2O_2 are added to the polymers. Hydrozen peroxide (H_2O_2) provides stability to the polymers by dissociating into two (OH) ions. This is shown in Fig. 7.3.

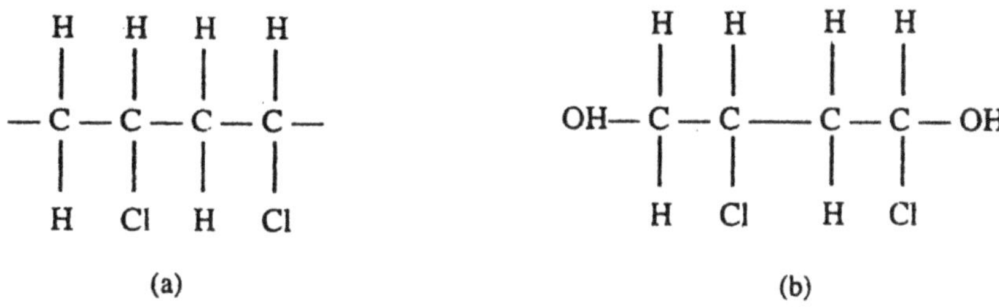

Fig. 7.3 *(a) Unstable polyvinyl chloride (b) Stable polyvinyl chloride.*

Fig. 7.3 shows the conversion of unstable polyvinyl chloride to stable polyvinyl chloride with the addition of H_2O_2 which dissociates into two (OH) ions and makes its link on either end of the chain.

7.1.1.1 *Copolymerization*

Copolymerization is a particular type of addition polymerization in which two or more chemically different monomers (compounds) are polymerized to form long chain molecules.

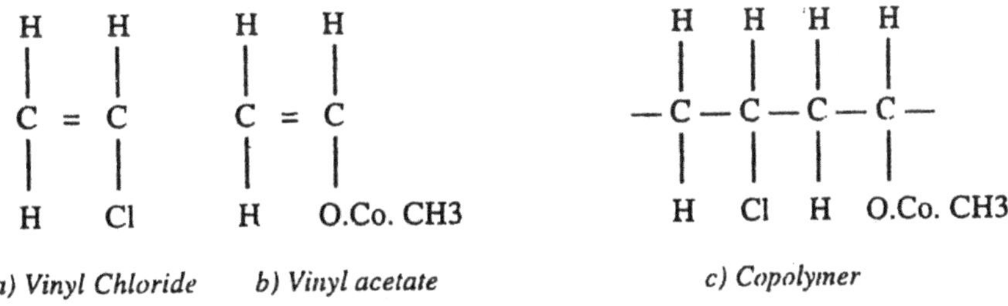

Fig. 7.4 *Copolymerization.*

Fig. 7.4 shows copolymerization process in which vinyl chloride monomer and vinyl acetate monomer are polymerized. During copolymerization process the double covalent bonds are broken and single covalent bonds are formed between carbon atoms. And the newly formed molecule as a result of copolymerization is known as copolymer of vinyl chloride and vinyl acetate. The process of copolymerization is widely employed in the field of manufacturing synthetic rubbers.

7.1.2 Condensation Polymerization

Condensation polymerization is the process in which two or more chemically different monomers are polymerized to form a cross link of linear polymer along with a by product such as water or ammonia. This process also occurs in unsaturated organic compounds and requines suitable conditions in the same way as that of addition polymerization.

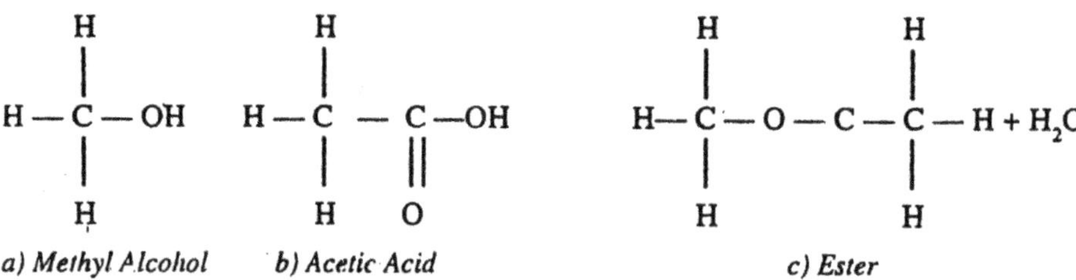

Fig. 7.5 *Condensation polymerization.*

Fig. 7.5 shows condensation polymerization process in which methyl alcohol monomer and acetic acid monomer are polymerized to form a long chain polymer known as ester with a by-product water.

7.1.3 Additions to Polymers

Before or during polymerization process certain additions are made to monomers in order to impart desired properties to the polymers. These additions are :

(a) Plasticizers : These organic compounds are oily in nature and of low molecular weight. These are added to prevent crystallization by keeping the chains separated from one another. These compounds give flexibility to the material and act as lubricants.

(b) Fillers : These are added to improve strength, dimensional stability and thermal resistance. The common filler materials are wood, asbestos, glass, fibres, mica and slate powder.

(c) Catalyst : These are added to expedite and complete the polymerization reaction.

(d) Initiators : These are added to initiate the polymerization reaction and to stabilise the end reaction of the molecular chains. The common initiator is H_2O_2.

(e) Dyes and Pigments : These are added to impart the desired colour to the finished polymers.

7.1.4 Electrical Behaviour of Polymers

All the polymers are poor conductors of electricity at room temperature. But they possess good insulation resistance and high dielectric constant. Hence polymers are widely used as insulating as well as dielectric materials.

7.2 PLASTICS (RESINS)

A plastic is an organic polymer which can be moulded into any desired shape and size with the help of heat, pressure or both. Plastics are most attractive organic materials for engineers because of their low specific gravities, ease of fabrication, resistance to solvent action, low thermal and electrical conductivities. The initial modulus of elasticity of plastics will be from 10^8 to 10^9 dynes/cm^2. Plastics in liquid form is known as resins.

The plastics are broadly classified as :

(a) Thermoplastics

(b) Thermosetting plastics

7.2.1 Thermoplastics

Thermoplastics are the polymers which are viscoelastic at intermediate temperatures and liquids at high temperatures i.e., their plasticity increases with increase in temperature. These plastics have linear molecules since they are produced by addition polymerization without any cross linking in structure. Hence thermoplastics can be mechanically deformed and softened at high temperatures and can be easily moulded or extruded due to the absence of cross links. On cooling to room temperature, they regain their original room temperature properties but they retain the shape and size into which they were moulded at high temperatures

Most common thermoplastics are poly-ethylenes (polythenes), polyamides (nylons), polystyrenes, polyvinyls, polypropylenes, acrylics, Teflon etc.

7.2.2 Thermosetting Plastics

Thermosetting plastics are the polymers which can not be softened once they are moulded, even at high temperatures i.e., their plasticity does not change with temperature. These plastics have cross linked molecules since they are produced by condensation polymerization. Hence these plastics can not be softened after moulding and cannot be moulded into any new shape due the presence of cross links. They do not deform on mechanical stressing. At very high temperatures, the cross links can be broken but at this temperature all the useful properties of the material are destroyed. This process is known as depolymerization or degradation of the polymers.

Most common thermosetting plastics are urea formaldehydes, phenolics, polyesters, epoxides etc.

7.2.3 Comparison between thermoplastics and thermosetting plastics

The following table gives the comparison between thermoplastic and thermosetting plastics.

Thermoplastics	Thermosetting Plastics
1. They can be repeatedly softened by heat and hardened by cooling.	1. Once hardened and set, they do not soften with the application of heat.
2. They are comparatively softer and less strong.	2. They are more stronger and harder than thermoplastics.
3. They are produced by addition polymerization.	3. They are produced by condensation polymerization.
4. They are usually supplied as granular material.	4. They are usually supplied in monomeric or partially polymerized form in which they are either liquids or semi solids.

7.3 MOULDING METHODS OF PLASTICS

The various moulding methods with which components are produced from plastics are as follows :

    (a) Compression Moulding

    (b) Transfer Moulding

    (c) Injection Moulding

    (d) Extrusion Moulding

    (e) Blow Moulding

    (f) Calendering

    (g) Thermoforming

    (h) Casting

7.3.1 Compression Moulding

Compression moulding is usually employed for thermosetting plastics.

A typical mould employed for compression moulding is shown in Fig. 7.6. The mould consist of two halves - the upper and the lower halves or the male and the female. Usually the lower half contains a cavity and the upper half consists a projection which fits into the mould cavity when the mould is closed. The gap between the projected upper half and the cavity in the lower half gives the shape of the moulded component.

In compression moulding, the thermosetting material which is partially polymerized and made in the form of pellets is placed in the heated mould. As the mould closes down under pressure (obtained by a hydraulic press) the material is squeezed or compressed between the two halves of the mould. The combined effect of heat and pressure causes the material to flow in the mould cavity. Under the influence of heat, the compacted mass gets cured and hardened to shape. The mould remains closed until the curing is complete. The time required for curing varies from 1 to 15 minutes depending on the size and thickness of the moulded component. Once curing is completed, the mould is opened to get the moulded component.

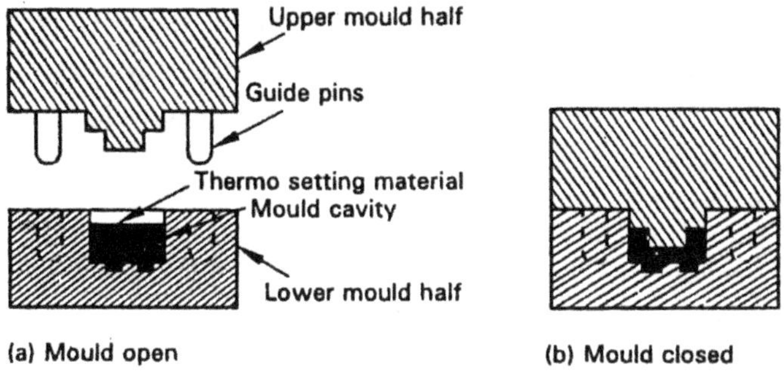

Fig. 7.6 *Compression moulding (a) Mould open (b) Mould closed.*

The mould temperature and pressure of thermosetting material varies from 150 °C to 180 °C and from 135 – 535 Kg/cm^2 depending upon the type of material.

7.3.2 Transfer Moulding

Transfer moulding is similar to compression moulding except that in transfer moulding, the moulding material (thermosetting pellets) is loaded in a separate loading chamber which is called transfer chamber. After loading the moulding material in the heated transfer chamber, the plunger is forced, which makes the fluid material to flow through a sprue into the mould cavity where final curing takes place.

Complex parts with thin sections can be made more easily by transfer moulding. Transfer moulding is used for producing intricate, delicate and dimensionally accurate components.

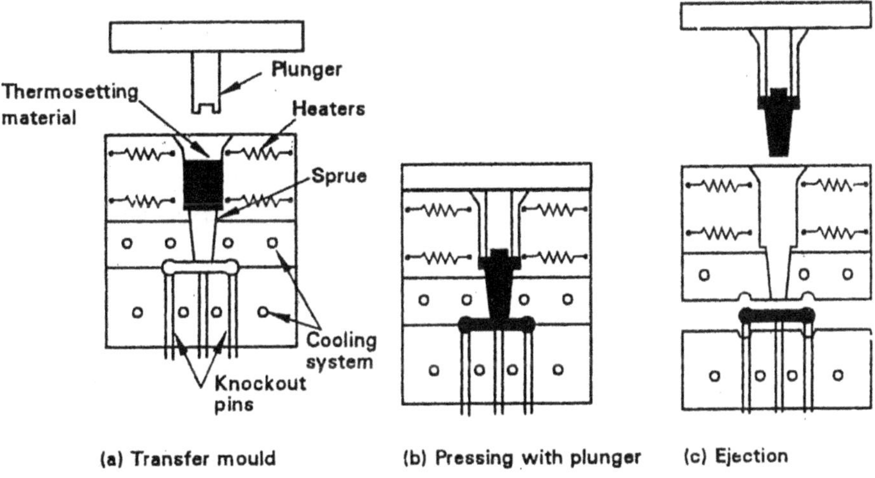

Fig. 7.7 *Transfer moulding process (a) Transfer mould (b) Pressing with plunger (c) ejection.*

7.3.3 Injection Moulding

Injection moulding is usually employed for thermoplastic materials which are softened by heating and then reharden during cooling. The moulding material (thermoplastic) is fully polymerized before pouring into the mould since they undergo only a physical change during injection moulding. In injection moulding the material is heated to the soften state and then the soft material is injected into the mould and then allowed to cool to harden thereby the desired article is obtained.

Fig. 7.8 shows the injection moulding method for thermoplastic material. The mould used consists of two parts, one is movable and the other is stationary. The stationary part is fixed to the end of the cylinder while the movable part is opened or locked on to the stationary part.

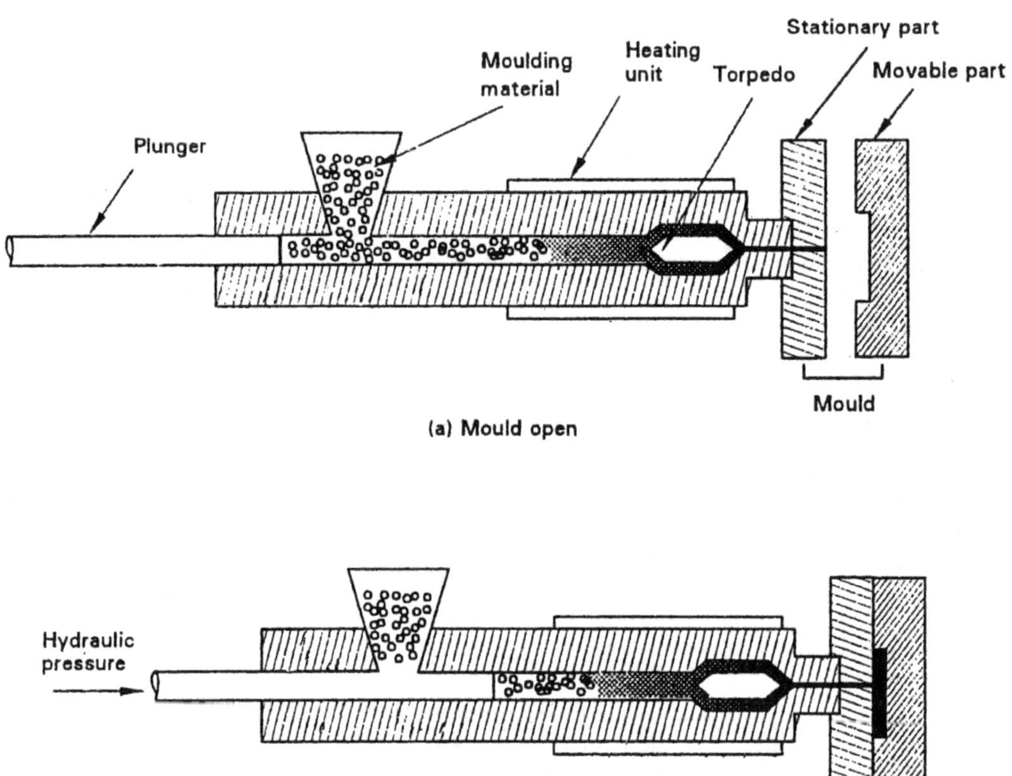

Fig. 7.8 Injection moulding method (a) Mould open (b) Mould closed and pressure is applied.

In injection moulding method, the moulding material in the form of granules or pellets is fed through the hopper to the cylinder where it gets softened. Then the mould is closed tightly and the plunger is moved forward by applying hydraulic pressure to push and inject the soften material through a nozzle into the water cooled mould. After the mould is filled it is allowed to cool and harden. Then the plunger return backwards and the mould is opened to eject the plastic articles.

A device called torpedo helps in spreading the moulding material uniformly around the walls of the hot cylinder thereby ensures uniform heat distribution.

Heating unit temperatures usually ranges from 150 °C to 300 °C and injection pressure ranges from 100 N/mm^2 to 150 N/mm^2.

The advantage of injection moulding is that mass production is possible which offset the higher capital cost.

7.3.4 Extrusion Moulding

Extrusion moulding is suitable for thermoplastic materials. Extrusion moulding involves feeding of moulding material from the hopper along the heating chamber into the heated die of the required shape with the help of a rotating screw.

In this process, the moulding material in the form of powder or granules is fed through the hopper into a heated cylinder. The moulding material is heated as it moves towards the die and gets softened (Fig. 7.9). A rotating screw is used for carrying the material through the cylinder and then to force the material which is softened by heat into the die. The die orifice controls the required shape. As the extruded material leaves the die, it is cooled rapidly enough to avoid deshaping and carried on a conveyor arrangement.

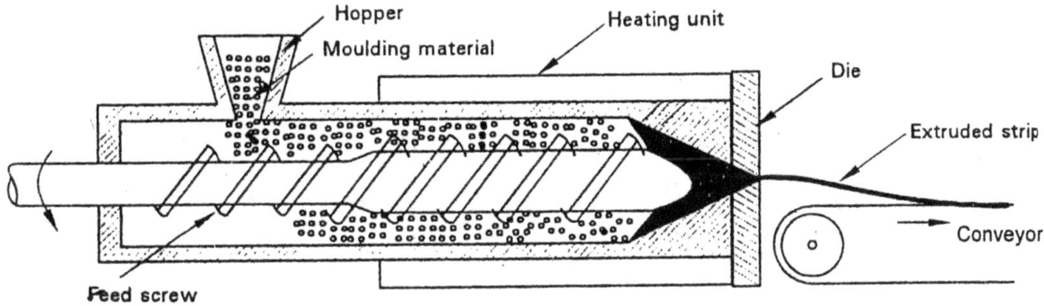

Fig. 7.9 Extrusion Moulding Method.

Extrusion moulding is employed for producing rods, tubes, film decorative strips and also for coating wires and cables with thermoplastics.

7.3.5 Blow Moulding

Blow moulding is also suitable for thermoplastic materials and produces hollow articles like containers, soft drink bottles etc.

Fig. 7.10 shows various steps involved in a blow moulding process. The blow moulding process starts with the production of a tube made of thermoplastic material. This tube is called as 'Parison'. The parison required for blow moulding can be made either by injection or by extrusion moulding. The parison is then kept in position inside a two piece mould and the mould is closed. Compressed air is then blown into the parison and the parison inflates like a balloon and goes on expanding until it comes in contact with the relatively cold walls of the mould and takes the shape of the mould under pressure. The mould is allowed cool and then opened to release the article.

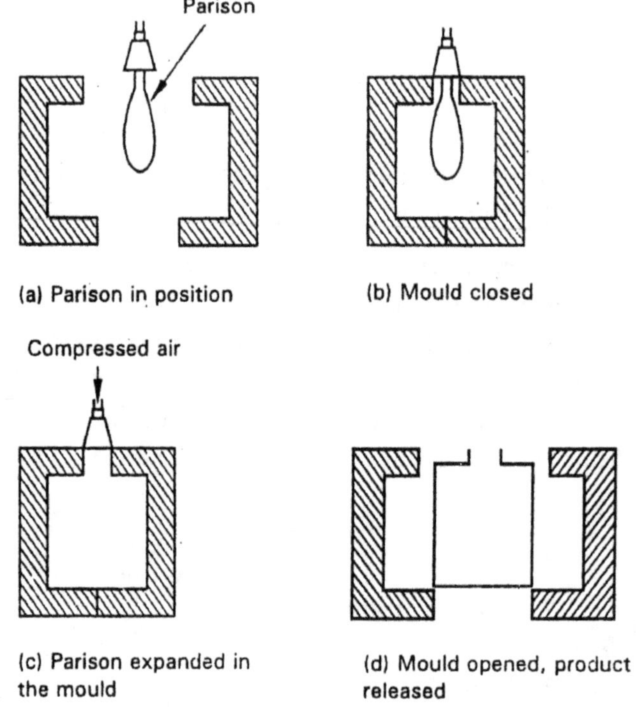

Fig. 7.10 *Blow moulding process*

(a) Parsion in position (b) Mould closed (c) Parsion expanded in the mould (d) Mould opened, product released.

7.3.6 Calendering

Calendering produces sheets and films from plastic materials such as polyvinyl chloride, polyethylene etc.

In calendering, first the plastic material is heated for a short time to produce a rough sheet wrich is then fed to a series of hot rolls. The gap between the rolls control the thickness of the sheet. The thickness of the sheet is gradually reduced and the final thickness of the sheet is controlled by the gap between the last pair of rolls and the sheet emerging from the rolls is cooled on a cold roll.

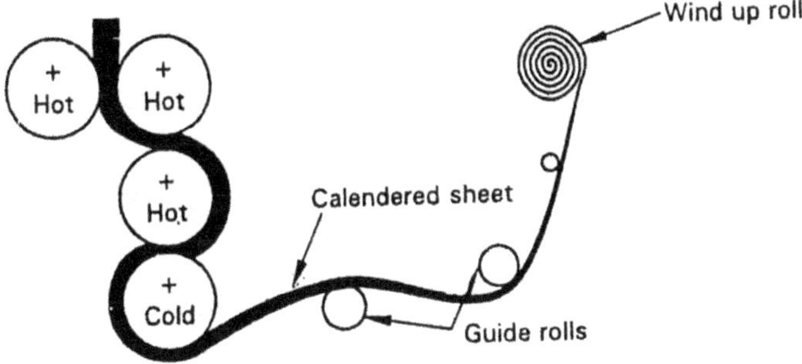

Fig. 7.11 *Calendering process.*

7.3.7 Thermoforming

Thermoforming involves heating a thin plastic sheet until it is softened and then forming it either mould halves or by vacuum.

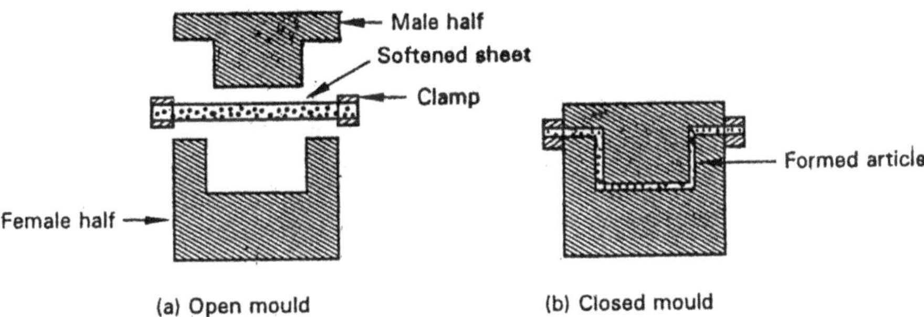

Fig. 7.12 *Thermoforming by matching mould halves (a) open mould (b) closed mould.*

In thermoforming by matching mould halves technique, first the heated and softened plastic sheet is held in position inside the mould. Then the two halves of the mould i.e., the female half and the male half is matched and the mould is closed (Fig. 7.12). Thus the softened sheet assumes the shape of the mould. The mould is allowed to cool and then the mould is opened to release the shaped article.

In thermoforming by vacuum, a thin plastic sheet is clamped to the mould and heated to soften. Then vacuum is applied to the mould (Fig. 7.13). This causes the softened plastic sheet to get pulled rapidly against the mould thereby acquires the shape of the mould.

The equipments and moulds used for vacuum forming are less costly and any desired shape can be obtained by vacuum forming.

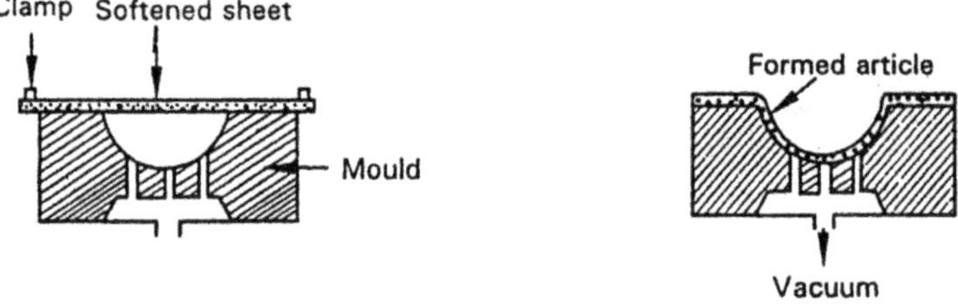

(a) Softened sheet is placed above the mould   (b) Vacuum is applied to the mould

Fig. 7.13 *Thermoforming by vacuum*

(a) softened sheet is placed above the mould (b) vacuum is applied to the mould.

7.3.8 Casting

The casting of plastics is similar to metal casting. The casting of plastics involves pouring liquid resin into the mould and allow it to solidify so that the liquid acquires the shape of the mould and gets hardened after cooling. Rubber, plaster of paris, wood or metal can be used as materials to make moulds for plastics. Thermosetting plastics such as epoxy, phenolics and polyesters and thermoplastics such as acrylic, polyethylene and urethane can be cast.

The moulds used are comparatively low in cost and since a longer time is required for hardening to take place, the process is employed for small scale production.

7.4 LAMINATING PLASTICS

In laminating plastics, a variety of materials such as sheets of paper, asbestos, fabric and wood are first impregnated or coated with resin and then bonded together by applying heat and pressure to form commercial materials. Depending upon the pressure required for compacting, laminations are classified into three types such as contact pressure laminates, low pressure laminates and high pressure laminates.

Contact pressure laminates are produced by applying pressure in the range of 0 and 17.5 N/cm^2. The resin used are epoxies and polyesters.

Low pressure laminates are produced by applying the pressure in the range of below 275 N/cm^2. This process is used for producing plywoods and sandwich type construction.

High pressure laminates are produced by applying the pressure in the range of 800 to 1400 N/cm^2. The resins used are phenolics, urea and the products are tubes, rods, sheets and special shapes.

7.5 JOINING OF PLASTICS

Plastics can be joined by several methods. The simplest method is by using mechanical fasteners such as screws, rivets, pins, nuts etc. For mechanical fastening, the plastic should be strong enough to withstand the strain of fastener insertion and subsequent high stress around the fastener.

Plastics can also be joined by using a solvent. This method is used for thermoplastics only. In this method the plastics are soften by applying a solvent and then clamped and pressed together to form the joint.

Frequently plastics are joined by welding. The various welding methods are heated tool welding, hot gas welding, induction welding, ultrasonic welding and friction welding.

In heated tool welding, the plastic parts to be joined are made to contact with a hot plate thereby the parts are softened and then the hot plate is removed and slight pressure is applied at the joint. Then the joint is cooled.

In hot gas welding, the parts to be joined and a filler rod (thermoplastic) are softened by heating with hot inert gas and then the parts are pushed together.

In induction welding, the two thermoplastic pieces to be joined are pressed together with a metal wire and then the metal wire is heated up by induction heating which in turn melts the plastic. Then by applying pressure on the joint, the two pieces are joined.

In ultrasonic welding, the parts to be joined are placed together and then ultrasonic vibrations are transmitted to them which vibrates the parts against each other thereby the parts are heated and fused together. This process is used for thermoplastics only.

In friction welding, one part is held stationery while the other is rotated and forced against the stationary part. The parts thereby fuse together due to the heat generated by friction.

7.6 MACHINING OF PLASTICS

Usually plastic components require little or no machining. In case machining is required to get close dimensional tolerances, machining may be performed by conventional machine tools and with ordinary tools. The main problem during machining is the generation of heat which softens and distort the work piece. This can overcome by using coolants such as air jet, vapour mist or 10 percent solution of water in soluble oil and also keeping the tool very sharp. Sharp edged cutting tools made with high speed tool steel or cemented carbide can work satisfactorily on plastics.

7.7 MODEL QUESTIONS

1. Differentiate between thermo plastics and thermosetting plastics.

2. List out various moulding methods and explain with neat sketch the compression moulding of plastics.

3. Describe with a neat sketch the transfer moulding process of plastics.

4. Differentiate between injection moulding and extrusion moulding.

5. With neat sketches explain blow moulding process of plastics.

6. Explain the following:
   - (i) Calendering
   - (ii) Casting of plastics

7. Explain thermoforming of plastics with necessary sketches.

8. Briefly discuss the various methods employed for joining of plastics.

9. Write a note on machining of plastics.

10. With neat sketch explain the injection moulding process of plastics.

8 POWDER METALLURGY

Like every modern technology powder metallurgy started as an art. It is the oldest kind of metallurgy (3000 B.C) and practiced before man was able to develop temperatures sufficiently high to melt metals.

Powder metallurgy is defined as the art and science of producing fine metal powders and objects finished or semi finished shape from individual, mixed or alloyed metal powders with or without the inclusion of non metallic constituents.

To produce metallic parts, the powder metallurgy process consists the following steps :

    (i)   Production of metal powders

    (ii)   Powder conditioning

    (iii) Compaction of metal powders

    (iv) Sintering

8.1 PRODUCTION OF METAL POWDERS

The methods for the production of metal powders are classified as :

(A)   **Mechanical processes**

    (i)   Machining

    (ii)   Crushing

    (iii) Milling

 (iv) Shotting

 (v) Graining

 (vi) Automization

(B) Physico-Chemical Processes

 (i) Condensation

 (ii) Thermal decomposition

 (iii) Reduction

 (iv) Electro deposition

 (v) Precipitation from aqueous solution

By using one or more of these methods, any metal can be practically made into powder. The selection of a particular method much depend on the type of raw materials available, the desired properties, type of application etc.

8.1.1 Mechanical Processes

Mechanical disintegration process, for the production of powders, are widely employed in powder metallurgy. Of the six methods mentioned in this classification, the first three involves disintegration of solid metals and the next three involves disintegration of liquid metals in order to produce metal powders.

8.1.1.1 *Machining*

In this method filings, turnings, chips etc., are produced and are subsequently pulverized by crushing and milling. The powder produced in this method will be coarse in size and irregular in shape. Hence this method is used for a few special cases such as production of magnesium powder (the explosiveness and malleability of this powder prohibits use of other methods), silver solders and dental alloys.

8.1.1.2 *Crushing*

Crushing is employed for disintegration of oxides and brittle materials. The brittle solids are crushed by using hammers, Jaw crushers or gyratory crushers. The powder particles produced by crushing possess angular shape and coarse in size which are subsequently comminuted by milling to obtain the required fineness.

8.1.1.3 *Milling*

Both brittle and ductile metals can be pulverized by milling. Milling involves the application of impact force on the materials to be pulverized. This milling action can be carried out by using ball mills, rod mill, impact mill, disk mill, vortex mill etc. During milling, work hardening, oxidation of the fine powder, particle welding and agglomeration occurs. All these are most undesirable for subsequent processing such as compaction and sintering. But, these can be eliminated by annealing the powders in a reducing atmosphere.

8.1.1.4 *Shotting*

Shotting involves the disintegration of molten metals. In this method, the molten metal is poured on a vibrating screen which disintegrates the molten metal into a large number of droplets. These droplets are allowed to solidify either in air or neutral gas atmosphere. The size and character of the resultant shot (particle) depends on the temperature of molten metal, size of openings in the screen and frequency of vibrations of the screen. Shotting produces spherical particles. Copper, brass, aluminium, tin, zinc, silver, lead, nickle, gold etc., metal powders are usually produced by shotting.

8.1.1.5 *Graining*

Graining involves the same procedure as shotting, but the only difference is that the solidification of molten metal droplets is carried out in water.

The powders produced by shotting and graining methods are coarse in size and hence other pulverizing methods are used for further reduction of size.

8.1.1.6 *Atomization*

Atomization involves mechanical disintegration of molten metal stream into fine particles by means of a jet compressed air, inert gases or water.

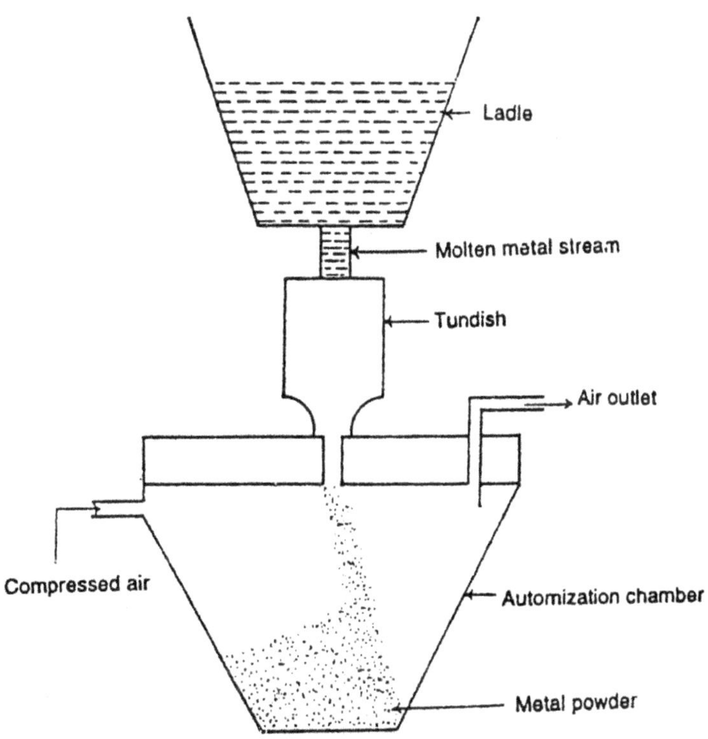

Fig. 8.1 Atomization Process.

As shown in Fig. 8.1, in this method the molten metal is forced through a small orifice and is disintegrated by a powerful jet of compressed air into small particles. These small particles are then allowed to solidify. Atomization produces generally spherical shaped particles. Since this process produces very small particles which are erroneously called atoms, the process is named as atomization.

By varying the temperature of the metal, pressure and temperature of the atomizing gas, rate of metal flow through the orifice and the design of the orifice, a wide range of particle size distribution may be obtained.

Since the methods of shotting, graining and atomization involves molten metals, they are specially suitable for metals of low and moderate melting points such as tin, lead, zinc, cadmium and aluminium.

8.1.2 Physico-Chemical Processes

Of the five methods mentioned in this classification, condensation and thermal decomposition are regarded as physical methods, reduction and precipitation from aqueous solution are regarded as chemical methods whereas electro deposition is regarded as an electro chemical method.

8.1.2.1 *Condensation*

In this method, metals are boiled to produce metal vapours which are then condensed to obtain metal powders. This process is usually applied to volatile metals such as zinc, magnesium and cadmium which can be easily transformed to their vapours. Condensation produces very fine powders with nearly spherical shape.

8.1.2.2 *Thermal Decomposition*

Metal powders of metals such as Fe, Ni, Mg, Mo and Cr can be produced by thermal decomposition of their respective carbonyls. Metal carbonyls are produced by passing carbon monoxide gas at a suitable temperature and pressure. The thermal decomposition of various metal carbonyls at a suitable temperature and pressure are shown below :

$$Fe\,(CO)_5 \;\rightarrow\; Fe + 5\,CO$$

$$NI\,(CO)_4 \;\rightarrow\; Ni + 4\,CO$$

$$Mo\,(CO)_6 \;\rightarrow\; Mo + 6\,CO$$

$$W\,(CO)_6 \;\rightarrow\; W + 6\,CO$$

This thermal decomposition gives metal powder of high purity and the carbon monoxide gas which can be recycled to produce metal carbonyls.

8.1.2.3 *Reduction*

Reduction is probably the oldest method of producing metal powders. In this process, the reduction of compounds like metal oxides, oxalate or halides is carried out by reducing agent (either solid or gas) such as carbon, hydrogen and carbon monoxide and the reduced powder is subsequently ground. This method is most suitable to produce powders of metals such as Fe, Cu, Ni, W, Mo. Infact this is only the practical method available for producing powders of refractory metals such as tungsten and molybdenum. Reduction produces very fine powder particles and the shape of particles is irregular.

8.1.2.4 *Electro Deposition*

Extremely pure metal powders can be produced by electro deposition. This process is similar to electro plating. In this, during electrolysis, the metal powder can be directly deposited on the cathode by controlling the current density, temperature, circulation of the electrolyte, composition of the electrolyte. The powder deposited on the cathode can be removed, washed and dried to obtain pure powders.

Electro deposition is mainly employed for the commercial production of metal powders such as copper, beryllium, iron, zinc, tin, nickle, cadmium, antimony, silver and lead.

Electro deposition produces powders having dendritic shape and purity of the powders is extremely high and may exceed 99.9 percent.

8.1.2.5 *Precipitation from Aqueous Solution*

This process is based on the principle that a metal may be precipitated from its aqueous solution by adding to it a less noble metal which is higher in the electromotive series. For example silver powder is produced from silver nitrate solution by adding a less noble metal such as copper or iron. Similarly tin powder is produced from stannous chloride solution by adding a less noble metal such as zinc.

This method is extensively used for producing copper powders. Copper powder is produced from copper sulphate solution by adding the less noble metal such as iron.

Precipitation from aqueous solution produces powders having dendritic shape with high purity.

8.2 POWDER CONDITIONING

The metal powders produced by the above processes, may not possess favourable physical or chemical characteristics for subsequent processing like compaction and sintering. Therefore the powder has to be conditioned before used for compaction and sintering. Powder conditioning consists of annealing treatment and blending or mixing.

8.2.1 Annealing Treatment

Annealing of powders is usually carried in reducing atmosphere or in vacuum. Annealing eliminates work hardening effect, reduces the oxide content and impurity level and alters the apparent density (apparent density of a powder is defined as the mass per unit volume of loose or unpacked powder). High temperature annealing increases the apparent density of powder and reduces the pressure requirements whereas low temperature annealing decreases the apparent density of powder and increases pressure requirements during compaction.

During annealing treatment the powder particles get sintered and form a spongy mass. Hence the powder has to be pulverized and screened prior to its further use. Pulverizing should be done carefully so that work hardening does not occur.

8.2.2 Blending or Mixing

Blending is an operation of thorough intermingling of different powders of the same composition or various grades of the same powders whereas mixing refers to the thorough intermingling of powders of more than one material. Alloying powders, lubricants and volatilizing agents to give desired amount of porosity are added to the blended powders during mixing.

Mixing is carried mainly to produce uniform distribution of particles so that the mixture will have the same chemical composition throughout the mixture volume. The time for mixing may vary from a few minutes to 24 hours or even several days depending upon the results desired. Overmixing decreases particle size and causes work hardening of particles, hence it is avoided.

Various types of blenders and mills are employed for mixing. Usually ball mills or rod mills are used for mixing hard metals such as carbides otherwise double cone mixers are most popularly employed. In cone mixers rolling action is provided to the powder mass and continuous dividing and recombining of powder mass are made possible at each revolution.

In order to remove any unwanted foreign substances or balled up aggregates of powder and lubricant, the mixed or blended powders are usually screened after the mixing operation.

8.3 COMPACTION OF METAL POWDERS

The main purpose of compaction is to produce metal compacts of desired shape from metal powders. These compacts should have sufficient strength to withstand ejection from the tools and subsequent handling upto the completion of sintering.

The compacting methods are classified into the following groups :
  (i)   Pressure less shaping technique
  (ii) Cold pressure shaping technique
  (iii) Pressure shaping technique with heat

8.3.1 Pressure less Shaping Technique

In pressure less shaping methods, the powder compacts are produced without the application of any external pressure. The various methods in this group are as follows :

    (i)   Loose sintering

    (ii)   Slip casting

    (iii) Slurry casting

8.3.1.1 *Loose sintering or Loose shaping or Gravity sintering or Pressure less sintering :*

In this method, the metal powder is poured or vibrated mechanically into a mould which is having a negative impression of the object and then heated to the sintering temperature to form the powder compacts. This method produces powder compacts with porosities varying from 50 to 90 percent by volume.

    The application of this method are:

    (i)   To produce highly porous filter materials such as bronze, monel, stainless steel, tungsten etc.

    (ii)   To produce extremely porous nickel sheets which are used as electrodes in fuel cells.

8.3.1.2 *Slip Casting*

Slip casting is extensively used for ceramics and only to a limited extent for metals. In this process, first a slip is prepared. The slip consists of powder particles (finer than 5 microns) in a liquid (mostly water) and a small amount of additives such as deflocculants (ammonium and sodium salts of alginic acids such as marex and keltex) which not only prevents the setting and segregation of the particles but also acts as binding agent to improve the green strength of the compact. The slip thus prepared is then poured into a mould made of fluid absorbing material such as plaster of paris. After partial drying in the mould (with the absorption of fluid by plaster of paris), the slip casting is removed from the mould and then dried in an oven. Now the compact i.e., slip casting is ready for sintering. Parts produced by this method will have high density than those of loose sintered components.

This technique is very useful for materials that are relatively incompressible by conventional die compaction. But the main disadvantage of this process is that it is a relatively slower process because long time is required for the fluid to be absorbed by the mould.

8.3.1.3 *Slurry Casting*

Slurry casting is similar to slip casting except that the slurry, which consists of metal powders in a suitable liquid (usually liquid resins or self solidifying liquids may be used) with various additives and binders, is poured into a mould made of plaster of paris and allowed to dry in the mould itself. Sometimes slow evaporation is used to remove the solvent in the slurry and cause drying of the slurry.

This process is also a slower process but it eliminates the expensive dies and equipments used in conventional compaction.

8.3.2 Cold Pressure Shaping Technique

The various cold pressure shaping methods are listed below :
- (i)　　Cold die compaction
- (ii)　　Isostatic pressing
- (iii)　Explosive forming
- (iv)　Powder rolling
- (v)　　Vibratory compacting

8.3.2.1 *Cold Die Compaction*

Compaction in metal dies is one of the most important method for shaping metal powders.

The compaction mechanism of metal powders in metal dies consists the following three stages :

- (i)　　In the first stage, particles are brought close together as a result of particle movement and rearrangement. There is no deformation of particles in this stage.
- (ii)　　The second stage attributes to plastic deformation and cold working. Since the applied pressure is more than the bulk yield stress of the material, plastic flow becomes homogeneous instead of local.
- (iii)　The third stage corresponds to bulk compression and involves fracture or fragmentation of brittle powders.

The basic components of the equipment required for compacting metal powders in metal dies are :

- (i)　　A mechanical or hydraulic mechanism for applying the desired pressure.

    Mechanical presses are available with pressure ratings of 10 to 150 tons and speeds of 6 to 150 strokes per minute. The special features of mechanical presses are high speed production rates, flexibility in design, simplicity and economy in operation. These presses require low investment and maintenance costs.

    Hydraulic presses have higher pressure ratings up to about 5000 tons but slower stroke speeds usually less than 20 strokes per minute. These presses are employed when high pressure is required especially for more complicated powder metal parts.

(ii)   A die of adequate strength having a cavity of the desired shape and size:

Dies are usually made of hardened, ground tool steels.

(iii)   Feeding devices for filling the die cavity with loose powder.

(iv)   Upper and lower punches of requisite strength and design to permit the application of pressure and to assist in ejection of the pressed compact and removal of the ejected compact from the die.

The punches are generally made of die steel which is heat treated to be slightly softer than the die, since the punches are usually easier to replace than the die. They must be perfectly aligned and very closely fitted.

(v)   Proper control to maintain the magnitude and rate of pressure application, the degree of movement and speed of the punches and in some cases the die and ejection of the pressed compact.

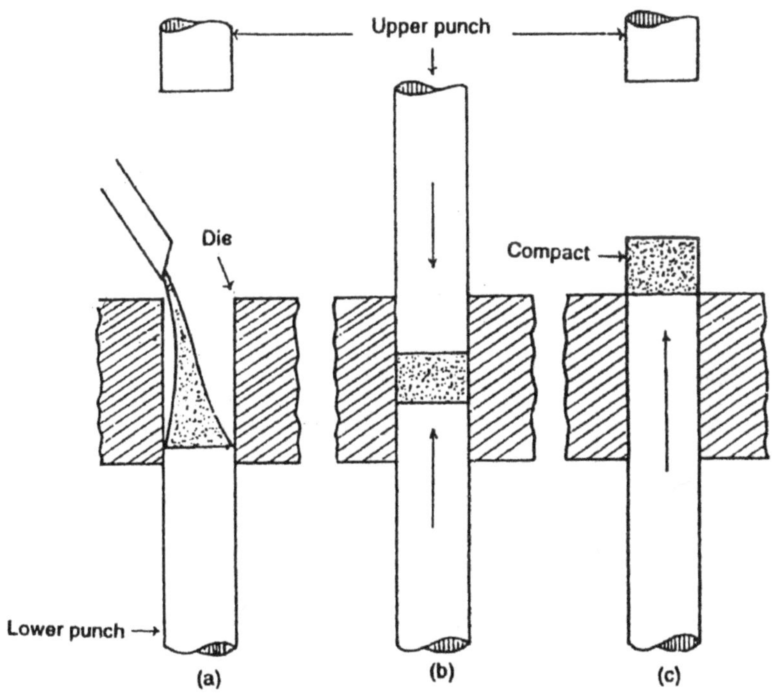

Fig. 8.2 *Cold die compaction method.*

The usual sequence of operations in die compaction are shown in Fig. 8.2. These are :

(a) Filling the die cavity with a definite volume of powder.

    (b) Application of the required pressure by movement of the upper and lower punches toward each other.

    (c) Ejection of the green compact by the lower punch.

Due to friction, a major loss of energy occurs during die compaction. The several types of friction that occur during die compaction are :

    (i)   Friction between moving punches and dies,

    (ii)   Friction between powder particles and the die wall,

    (iii)   Friction between the green compact and the die wall during ejection.

    (iv)   Friction between particles

Lubrication is usually done to overcome the frictional effects. These lubricants are of two types :

    (a) External lubricants : These are applied to the die walls, also called as die wall lubrication.

    (b) Internal lubricants : These are admixed with metal powders before compaction.

8.3.2.2 *Isostatic Pressing*

lsostatic pressing produces powder compacts using tightly sealed flexible moulds and applying a uniform pressure simultaneously and equally in all directions thereby achieving uniform density and strength.

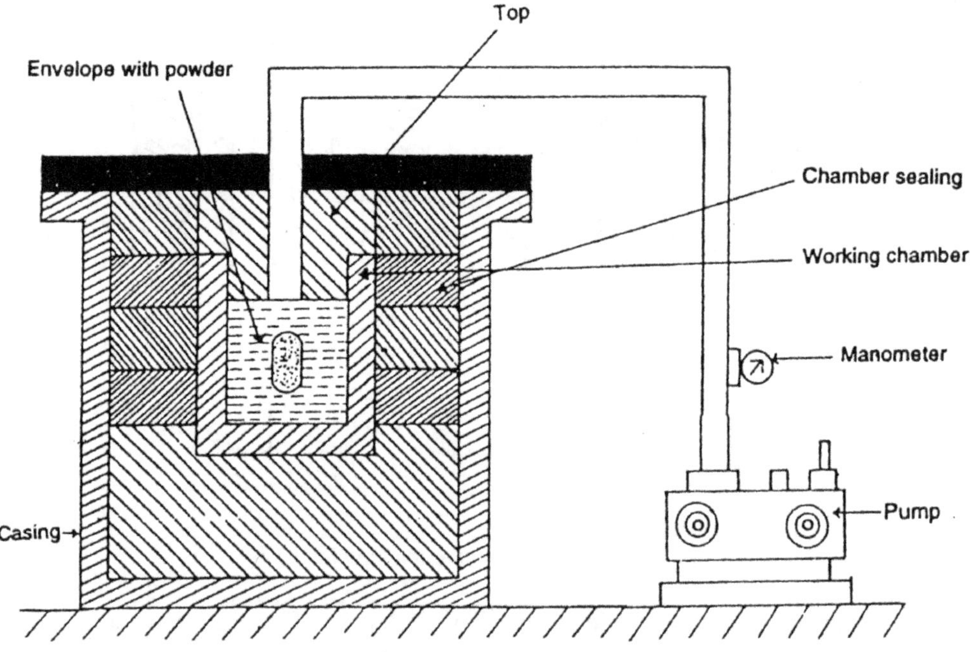

Fig. 8.3 Isostatic Pressing.

Fig. 8.3 shows a typical isostatic compaction. As shown in Fig. 8.3, the powder is loaded in a shaped flexible envelope for the production of desired shape and then tightly sealed. The flexible envelope is made from natural rubber, synthetic rubber, plastics or thin metal foils (The flexible envelope must be completely impervious to the pressurizing fluid and able to withstand internal pressure). The sealed flexible envelope is then placed in the sealed high strength pressure vessel (see Fig.) filled with oil. The fluid (oil) in the pressure vessel is pressurized to the required pressure with the help of a suitable high pressure pump. Application of this pressure causes the flexible envelope to impart the desired shape to the powder. After this, the pressure is then released and the pressed compact is taken out of the pressure chamber. The density of the compact depends on the pressure used in this process.

The main advantage of this process is the application of uniform pressure in all directions and hence the compact possess very high density near to the theoretical density. But the production rates are low because this is a batch process.

Isostatic compaction is employed to produce powder compacts of metal powders such as tungsten. molybdenum. niobium etc.

8.3.2.3 *Explosive Forming*

In this method the pressure generated by an explosive is used for the movement of punch or piston into the dies containing metal powder. The metal powder is well protected from the products of the explosives. The detonation (combustion) of low explosives such as smokeless powder and black powder produces pressures upto 3,00,000 pounds per square inch whereas high explosives such as dynamite, TNT (trinitro tolvene) produces pressures upto several millions of pounds per square inch.

Since very high pressures are involved in this process. the compacts produced by this process possess very high and uniform density of the order of 94 to 98 percent of the theoretical density.

8.3.2.4 *Powder Rolling*

The continuous compaction of metal powders by rolling has been first reported in Germany in 1950. In this process, as shown in Fig. 8.4, the rolls are set side by side so that the strip comes out vertically downward. The metal powder is poured from the hopper into the channel between the rotating rolls upto the required height on the roll surface. When the rolls start rotating, the powder is drawn into the gap between the rolls and is compacted into a strip.

At present powder rolling is mainly used for the production of metallic stripes and sheets.

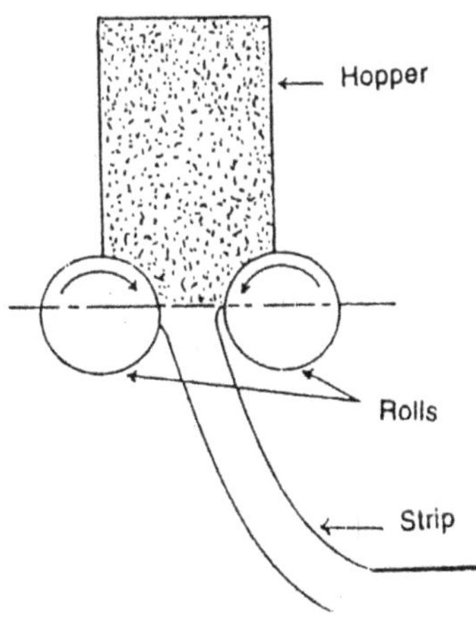

Fig. 8.4 Powder rolling.

8.3.2.5 *Vibratory Compaction*

In vibratory compaction method, both pressure and vibration are simultaneously applied during compaction of metal powders in rigid die. The application of vibration during compaction causes desired densification at a much lower pressures compared with ordinary die compaction. Both high frequency low amplitude and low frequency high amplitude vibrations have been applied successfully for achieving the given level of densification.

The major drawback of this process is the design of complex equipment for applying vibrations. This process is mainly suitable for producing very large and complex parts due to great reduction in the required pressure.

8.3.3 Pressure Shaping Technique with Heat

Pressure shaping with heat include the following process :
    (i)   Hot pressing
    (ii)   Powder (or) sinter forging
    (iii)   Hot rolling

8.3.3.1 *Hot Pressing*

To produce very large parts is a difficult task in cold die compaction because very high pressures are required. But in hot pressing where pressure and heat are simultaneously applied, large parts can be compacted to the required density with the application of only 10 to 20 percent of the cold die compaction pressure. The amount of pressure required for compaction decreases with increase in temperature.

In this process, the metal powder is poured into the die and then pressure and heat are simultaneously applied. Heating is usually done with the help of water cooled high frequency coil. Just by employing a pressure of about 100 kg/cm^2 and a temperature of two thirds of the melting point of the metal powder, compacts can be produced with about 90 percent of the theoretical density.

During the operations such as filling, heating and ejection, a protective atmosphere (usually argon or even vacuum) is maintained in order to prevent oxidation.

The major problem associated in this process is the rapid wear of the die. Therefore die material should be properly chosen. Usually hardened high speed dies are employed.

8.3.3.2 *Powder or Sinter Forging*

In this process, the powder compact, produced by conventional die compaction, isostatic compaction or slip casting, is heated to hot working temperature in a protective atmosphere. The heated compact is immediately transferred to closed die and the forging operation is done. To reduce oxidation, after forging, the component is removed from the die and is immediately either quenched or cooled in a protective atmosphere.

The components produced by this process possess 100 percent theoretical density.

8.3.3.3 *Hot Rolling*

The compacts produced by this process will have better properties than those produced by cold rolling. In this process, either hot rolls or hot powders are employed and the operation is similar to cold rolling. But the problem encountered in this process is to provide a protective atmosphere during rolling to prevent oxidation of the powder at high temperatures used in the process.

Hot rolling is most suited for molybdenum and tungsten based alloys which can be cold rolled with great difficult.

8.4 SINTERING

Due to the limited degree of interparticle bonding, the green compacts, produced by various compacting methods will be very fragile. Hence handling of these compacts between compacting and sintering is restricted to a minimum to avoid breakage. The most important step in powder metallurgy is the sintering process. Sintering imparts the required properties such as more strength, densification and dimensional control to the green compacts due to the formation of strong bond between the particles. After sintering the article can be used in almost all practical applications.

Sintering may be defined as "the heating of compacted aggregate of metal powders below the melting point of the base metal with or without the application of external pressure in order to transform it to a more dense material by interparticle bonding". Thus sintering is a progressive transition without melting from an agglomeration state to a massive state possessing the required physical and mechanical properties.

8.4.1 Sintering Process

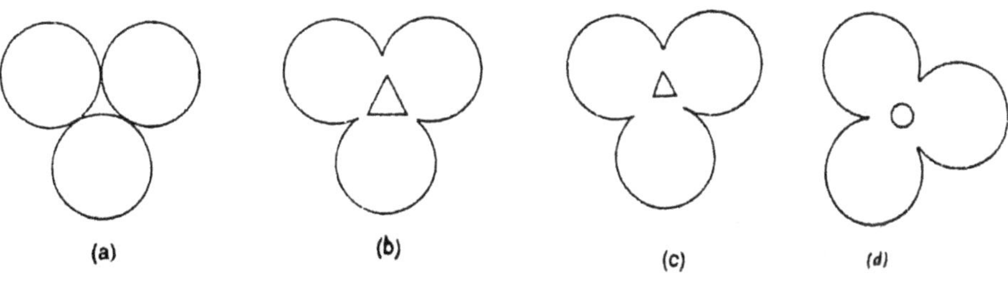

Fig. 8.5 Sintering Mechanism.

There are three stages involved in sintering process. These are :

(i)   In the first stage, as the green compact is heated, the sintering process starts with bonding among particles. Bonding involves diffusion of atoms where there is intimate contact between particles leading to the development of grain boundaries (Fig. 8.5 a). This interparticle bonding causes a large increase in strength and hardness even after short exposures to an elevated temperature.

(ii)   In the second stage, the newly formed bond areas called "necks" grow in size followed by pore rounding (Fig. 8.5 b)

(iii)   In the last stage, shrinkage of pore and its eventual elimination occur (Fig. 8.5 C & D). This stage is never complete because the temperatures and times required are too impractical.

During sintering, the formation of surface oxide films must be avoided since bonding between particles is much affected by surface oxide films. This can be achieved by providing a controlled protective atmosphere during sintering. Usually natural gas or propane or dissociated ammonia or vacuum is used as a protective atmosphere.

As a result of sintering, apart from interparticle bonding, the following changes also occur :

(a) dimensional changes

(b) chemical changes

(c) electrical properties changes

(d) phase changes

(e) relief of internal stresses

(f) alloying

In general, density, mechanical strength, ductility, electrical and thermal conductivity increase with increased amount of sintering.

8.4.2 Liquid Phase Sintering

Sintering in the presence of a liquid phase in multi component system is known as liquid phase sintering i.e., liquid persists during the entire sintering period.

Advantages :

(i)   Liquid phase sintering renders a density almost equal to the theoretical density. Due to liquid-solid interfacial tensions, almost all the pores are filled by the liquid.

(ii)   The rate of diffusion through the liquid phase is more and hence the rate of alloying will be greatly enhanced.

(iii) After liquid phase sintering, the compacts possess a more desirable metallurgical structure which gives excellent mechanical properties.

8.4.3 Sintering Furnace

Sintering furnace is required for heating the components during sintering. Electrical resistance type or gas or oil type of furnaces may be used for sintering. Since during sintering close control of temperature is very much essential, usually electric furnace is preferred where accurate control of temperature is possible.

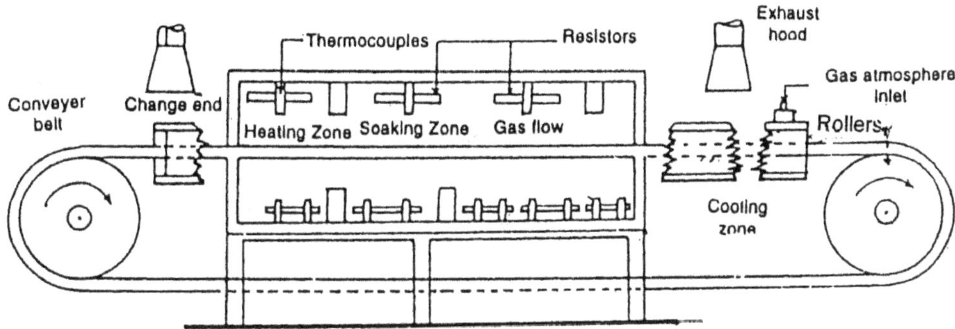

Fig. 8.6 *Continuous type sintering furnace.*

The essential parts of sintering furnace are :
   (i)   a burn off chamber for the removal of air or lubricant vapours,
   (ii) a controlled high temperature sintering zone, and
   (iii) a cooling and discharging zone

(a) Burn off Zone : The burn off zone is also called as the entrance zone which is an important part of the furnace where upto 900 °C temperature will be maintained. In this zone the green compacts are heated slowly, otherwise the entrapped air and lubricants will expand very rapidly so that the metal particles will be pushed apart. Before pushing the compacts into the high temperature zone i.e., sintering zone, all the lubricants used during compaction should be expelled as vapours along with furnace atmosphere.

(b) High Temperature Zone or Sintering Zone : In this zone, the actual sintering of the components occur. Sintering time and sintering temperature are the critical parameters, controlling of which causes good sintering. Proper sintering causes to get the desired density and strength in the powder metallurgy components otherwise results in poor strength and density.

Fig. 8.7 shows the curves of the mechanical strength versus sintering time at a given temperatures. As shown in Fig. 8.7, the degree of sintering increases with increase in sintering time but the effect is more pronounced with increase in temperature.

Table 8.1, shows the typical values of sintering temperature and sintering time for various metals.

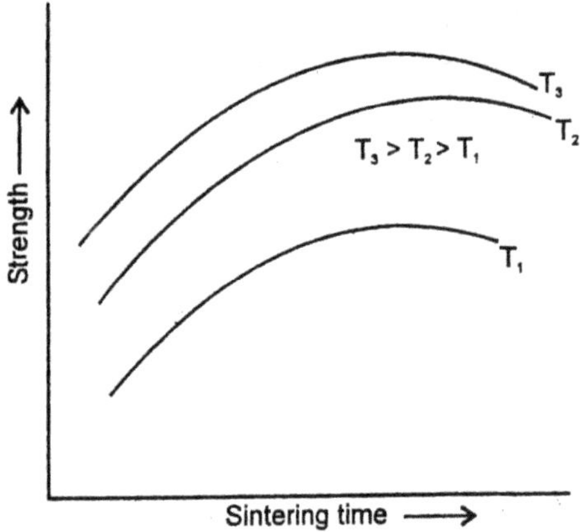

Fig. 8.7 Mechanical strength versus sintering time.

Metal	Sintering Temperature, °C	Sintering Time, minutes
Bronze	760-870	10-20
Brass	840-900	10-45
Copper	840-900	12-45
Nickel	1010-1150	30-45
Stainless steel	1090-1290	30-60
Molybdenum	2050	120 approx.
Tungsten	2340	480 approx.
Tantalum	2400	480 approx.

Table 8.1 *Typical sintering temperature and time for various metals :*

(c) Cooling Zone : The sintered parts from high temperature zone passes to the cooling zone of the sintering furnace where they are cooled so that oxidation does not occur on exposing to air. The cooling zone consists of two sections : These are :

(i)   *a short insulated cooling zone :* This allows the sintered components to cool from high sintering temperature to a low temperature. In this zone, the cooling is done at a slower rate in order to avoid thermal stresses.

(ii)   *a relatively long water jacketed cooling zone :* In this zone, the sintered components are cooled below 150°C to prevent oxidation of the components on exposing to air. Cooling in this zone is done for a long time i.e., very slow cooling rates are provided.

8.5 SECONDARY OPERATIONS AFTER SINTERING

The following secondary operatioris which will be carried after sintering impart better surface finish and properties to sintered parts.

(i) Coining : Coining is a process in which the sintered part is pressed in dies to reduce porosity and to increase density.

(ii) Sizing : During sintering dimensional changes occur in the parts. Sizing is the operation of pressing the part in dies to achieve the required dimensional accuracy. Sizing also improves surface finish and causes small changes in density.

(iii) Machining : Machining operation is usually carried on sintered part to provide under cuts, holes, threads etc., which can not be provided on the part during compaction because it becomes complicated design.

(iv) Impregnation : Impregnation operation is carried out to completely fill or close the voids with oil, grease or other lubricating oils. In this operation sintered parts are immersed in the lubricant and heated to a temperature of about 95 °C for 20 to 25 minutes and then removed. By doing this, the lubricant enters the pores by capillary action.

(v) Infiltration : Infiltration is the process of adding one metal to the pores of the sintered part. For example, copper blank is placed over sintered iron part (30 to 40 percent porosity) and heated to a temperature of 1150 °C in a furnace where the copper melts and infiltrates into the pores of the sintered iron part thereby producing a sintered part of 100 percent density. Infiltration also provides extra strength, hardness and toughness to the sintered part.

8.6 ADVANTAGES OF POWDER METALLURGY

Over conventional metal casting method, powder metallurgy has certain unique advantages. These are :

(i) High purity raw materials can be used and this purity can be maintained till the end of the process by controlling the fabricated steps.

(ii) Clean and smooth operations.

(iii) Close dimensional tolerances and smooth surfaces can be achieved.

(iv) No defects such as gas pockets, blow holes etc., are produced whereas these are common in metal casting.

(v) Various machining operations can be eliminated and this causes high production rates.

(vi)   New combinations can be achieved. For example metal and non metal can be mixed which is quite impossible by other methods.

(vii) Highly qualified or skilled personnel are not required.

(viii) Parts of any desired composition can be made. There is almost no need of referring to their equilibrium diagram.

8.7 LIMITATIONS OF POWDER METALLURGY

The following are some serious limitations of powder metallurgy which limits its application :

(i)   The cost of dies and equipment is very high and also rapid wear of dies occur during compaction. Hence it is only suitable for mass production and not for products less than 10,000 in quantity.

(ii)   This process is not used in some cases, due to the problem of storing the expensive powders with out deterioration.

(iii) Sintering parts possess lower tensile strengths, lower elongations and much lower impact strengths due to the presence of inherent porosity.

(iv)   Intricate shapes can not be manufactured easily with this process.

(v)   Oxidation of metal powders causes sintering difficult and results in inferior parts.

(vi)   Parts with theoretical density can not be manufactured.

(vii) Powders like magnesium, aluminium, titanium etc., causes some times explosions and fire hazards.

8.8 APPLICATIONS OF POWDER METALLURGY

Power metallurgy is utilized for manufacturing number of materials and products. Some of the important material and products produced by powder metallurgy are as follows :

(i)   Porous Metal Sheets

Metal powders such as copper, brass, Bronze and stainless steel can be rolled into porous sheets having controlled porosity. The following two processes are employed for making the porous sheets.

Gravity Sintering : This process is used for stainless steel. In this process, a uniform layer of powder is spread on ceramic trays. It is then sintered in an atmosphere of dissociated ammonia for 40 hours at high temperature. Then they are rolled to obtain uniform thickness and better surface finish. These porous sheets made of stainless steel powders are corrosion resistant and are used for gasoline, oils and chemical filters.

Rolling : This process is used for brass, bronze and copper. In this process, the metal powder is filled in a hopper and the powder is allowed to fall in between two metallic rollers, where the powder gets compressed into a sheet is then sintered and then rolled again to get the final size. This process results in uniform mechanical properties and controlled porosity.

(ii)   Porous Self Lubricating Bearings

By powder metallurgy technique, any desired degree of porosity can be obtained in metal products by controlling the particle size, its distribution and the pressure during compaction. Self lubricating bearings require 25-30 percent interconnected porosity. These bearings, after sintering, are subjected to oil impregnation so that the pores will be completely filled with oil. Hence during service, these bearings produce a constant supply of lubricant to the surface due to capillary actIon. In an assembly, where lubrication is not possible or oil splash is undesirable, self lubricating bearings can be used, for example in food industries.

Self lubricating bearings of bronze, copper, tin, graphite and iron are made by powder metallurgy.

(iii)   Gears and Pump Rotors

Gear and rotor for automobile oil pumps are manufactured by powder metallurgy. In this the iron powder is mixed with sufficient graphite so that it provides the desired carbon content to the product. It is compacted with a pressure of about 40 kg/cm^2 and then sintered in an inert atmosphere in electric furnace. The sintered products are then impregnated with oil. During service, under operating pressure, the oil seeps to the contact surface and when the load is removed, the oil will be reabsorbed.

(iv)   Electrical Contacts and Electrodes

Electrical contacts and resistance welding electrodes are made by powder metallurgy technique. Different combinations like tungsten-copper, tungsten-cobalt, tungsten-silver, silver-molybdenum and copper-nickel-tungsten have been developed to provide the characteristics of wear resistant, refractory and good electrical conductivity to the contact parts.

(v)   Magnet Materials

Small magnets known as alnico which is a mixture of aluminium, nickel and copper powders are produced by powder metallurgy technique. The magnets produced by powder metallurgy have better properties than the magnets produced by casting. They possess homogeneous structure, fine grain size and are free from internal defects.

(iv)   Diamond Impregnated Tools

Diamond impregnated tools which are used for cutting porcelain and glass are made from a mixture of iron powder and 30 percent diamond dust. This mixture is compacted at a pressure of 8 to 10 kg/cm^2 and then sintered at a temperature of 1000 °C. The diamond impregnated tool bit is generally welded or brazed to a steel.

(vii)   Metallic Coatings

Production of metallic coatings by powder metallurgy is a recent development. In this method a layer of metal powder (to be coated) is spread on a sheet of metal like iron, chromium or stainless steel and then heated in a hydrogen atmosphere for about 20 to 30 minutes at a desired temperature. After sintering the sheets are rolled. Iron can be coated with copper, nickel, silver, tungsten, molybdenum, chromium and cobalt powder followed by rolling into sheets.

(viii) Cemented Carbides

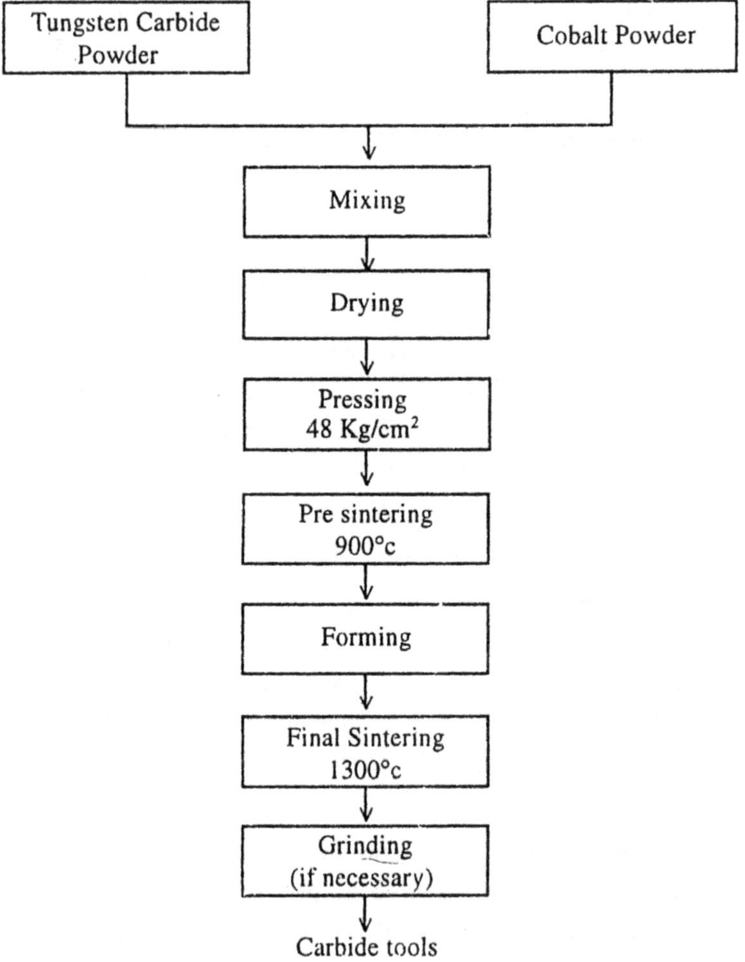

Fig. 8.8 Typical flow sheet showing the steps involved in the production of carbide tools.

For producing the cemented carbide cutting tools, the tungsten carbide and the cobalt powders are mixed in a mixing machine in a suitable proportion. Cobalt powder should be uniformly distributed in the tungsten carbide powder in order to obtain a strong bond. After mixing it is compacted with a pressure of 48 kg/cm^2 and then sintered. In this case, sintering is done in two stages, the pre sintering and final sintering. Pre sintering is done in a controlled hydrogen atmosphere at a temperature of about. 900 °c. At this sintering temperature, the tungsten carbide remains intact while the cobalt melts and forms liquid causing liquid phase sintering. In the pre sintered condition, the products can be cut, machined and ground to the final shape. Then the pre sintered compacts are subjected to final sintering at a temperature of 1300 °c for about 2 hours to develop full hardness. The compacts are cooled in the furnace gradually. After cooling if necessary these are ground. The cemented carbide tools are noted for high compressive strength, red hardness and wear resistance.

8.9 MODEL QUESTIONS

1. List out various methods employed for the production of metal powders.

2. Distinguish between shotting and Atomization

3. What are the advantages in blending or mixing metal powders?

4. Explain the following :
   - (i) Loose sintering
   - (ii) Slip casting
   - (iii) Slurry casting

5. With neat sketch explain the cold die compaction process.

6. Write notes on the following :
   - (i) Explosive forming
   - (ii) Powder rolling
   - (iii) Vibratory compacting

7. Explain the mechanism involved in sintering.

8. Explain the zones in sintering furnace.

9. What are the advantages and applications of powder metallurgy?

10. Explain three important applications of powder metallurgy.

11. With a flow sheet explain the steps involved in the production of carbide tools.

9 HEAT TREATMENT OF METALS

In the manufacturing process of many machine parts and tools, heat treatment plays an important role. It is possible to impart high mechanical properties to metals by heat treatment. Heat treatment may be defined as "A combination of heating and cooling operations, timed and applied to a metal or alloy in the solid state to obtain desirable conditions (for example relieved stresses) and properties (for example high strength, wear resistance, better machinability etc).

Heat treatment may be carried for the following purposes :

    (i)   to improve ductility

    (ii)   to relieve internal stresses

    (iii) to harden and strengthen metals

    (iv) to refine grain size

    (v)   to improve machinability

    (vi) to improve electrical and magnetic properties

9.1 HEAT TREATMENT CYCLE

Fig. 9.1 illustrates a simple heat treatment cycle. As shown in figure first the metal or alloy is heated to a definite temperature and held at that temperature for sufficient time to allow the necessary changes to occur and then cooled to room temperature. The rate of cooling depends on desired properties. The holding time depends on size of the object and generally provided at a rate of 2 to 3 minutes per millimeter of section thickness. Now before actually discussing the various heat treatment processes, let us study the T.T.T. diagrams and their importance in heat treatment of steels.

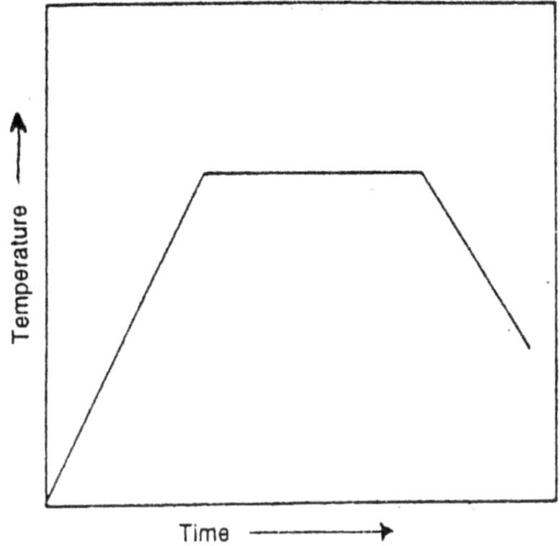

Fig. 9.1 *Simple heat treatment cycle.*

9.2 T.T.T. (TIME-TEMPERATURE-TRANSFORMATION) DIAGRAM

T.T.T. diagram is also called as S-curve, C-curve, Bain's curve or iso-thermal transformation Diagram. This diagram shows the relationship between temperature and time for the transformation of austenite when the transformation occurs at constant temperature (isothermal).

9.2.1 Importance of T.T.T. diagram

The iron-iron carbide diagram shows the phases that form under equilibrium conditions. But many heat treatment processes involves non-equilibrium rates of cooling. Under non equilibrium conditions the iron-iron carbide diagram has little value. For example at very high rates of cooling (quenching), a metastable phase called martensite is formed in steels. This martensite phase has no place on the iron-iron carbide diagram. Therefore the use of iron-iron carbide diagram is only to fix the austenitizing temperature. T.T.T. diagram is the main source of information for the decomposition of austenite under non equilibrium conditions.

9.2.2 Steps to construct T .T .T. diagram

The following are the steps involved in the construction of T.T.T. diagram.

Step 1　　　Prepare a large number of small specimens cut from the same steel bar.

Step 2　　　These specimens are then heated by placing in a furnace or molten salt bath which is maintained just above the austenitizing temperature. The specimens are held at this temperature for a long enough time to become completely austenite.

Step 3　　　The specimens are then placed in a furnace or molten salt bath which is maintained at a constant desired reaction temperature below austenitizing temperature for example 350 °C.

Step 4　　　After a given specimen is allowed to react isothermally for a certain time, it is then quenched in cold water or iced brine. Generally the first specimen is allowed to react isothermally for 2 seconds, second specimen for 4 seconds, third specimen for 8 seconds and so on upto say 10 hours. The more the time is given to a specimen to react isothermally, the more pearlite is formed.

Step 5　　　After cooling the specimens are checked for hardness and carried the metallographic examination. The result is the reaction curve which is shown in Fig. 9.2.

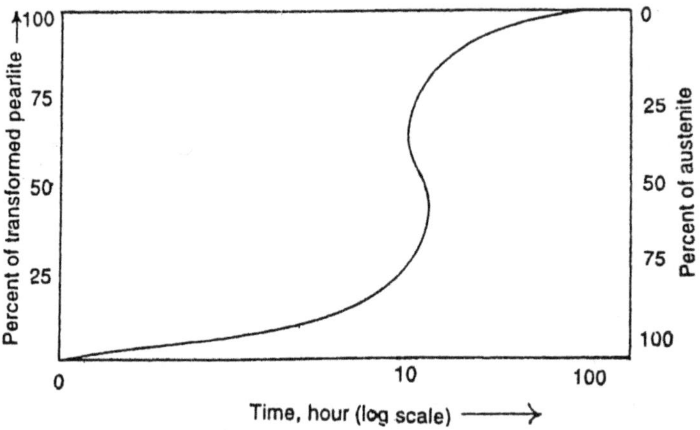

Fig. 9.2 *Reaction curve.*

Fig. 9.2 indicates that the transformation of austenite to pearlite is not linear and the rate of transformation is very slow initially then it increases rapidly and finally it slows down toward the end.

Step 6   The above steps are repeated at different sub critical temperatures to obtain a series of reaction curves over the whole range of austenite instability for a given steel. The result is the T.T.T. diagram for that steel.

From each reaction curve (for example at 350 °C) two points namely the time for the beginning and the time for the end of transformation may be plotted (Fig. 9.3)

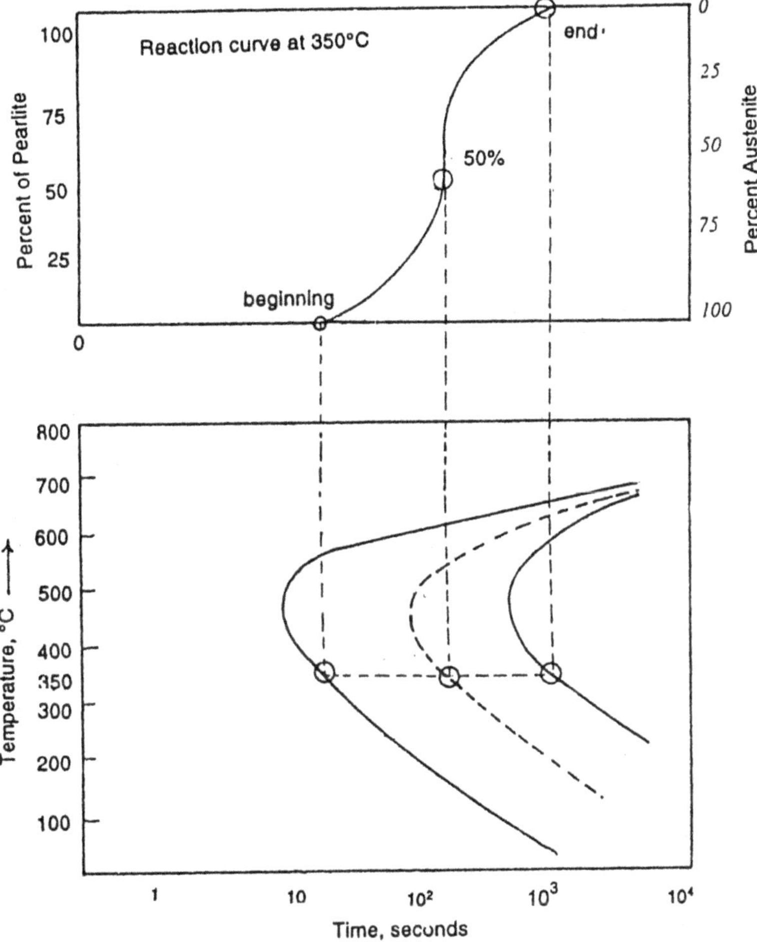

Fig. 9.3 *Construction of T.T.T. diagram.*

9.2.3 T. T .T. Diagram for an eutectoid steel

As shown in Fig. 9.4, austenite is stable above A_1 temperature (723 °C) line. Below this line austenite is unstable i.e., it can transform into different products such as pearlite, bainite or martensite.

The transformation product above the nose region (550 °C) is pearlite consisting of lamellar structure of alternate layers of ferrite and cementite. At a temperature just below A_1

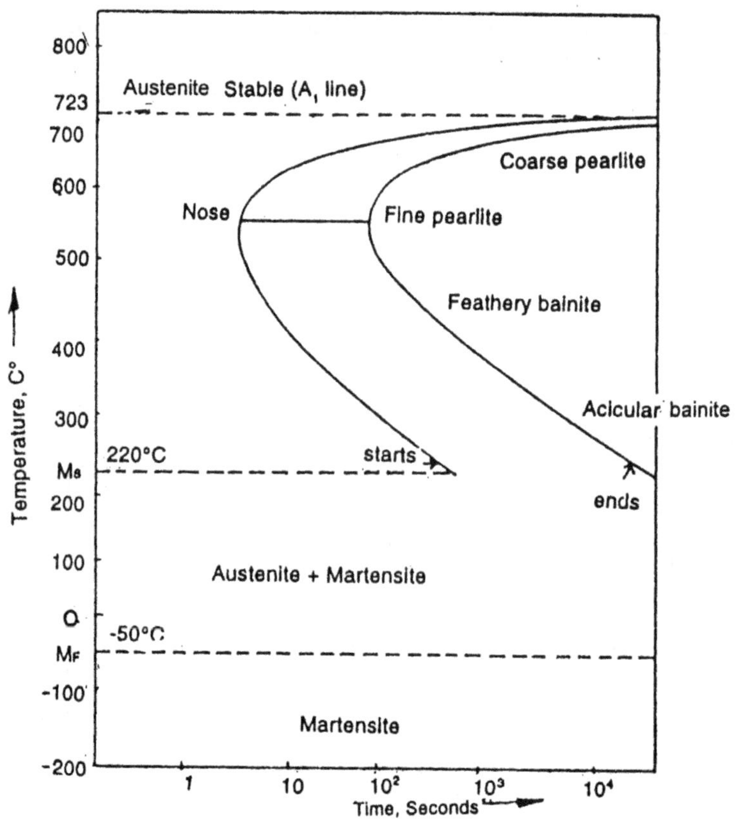

Fig. 9.4 *T.T.T Diagram for a plain carbon steel of eutectoid composition.*

line, nucleation of cementite from austenite will be very slow, but since diffusion and growth of nuclei will be maximum, the result is the formation of coarse pearlite. But in the region just above nose (550 °C), rapid transformation causes the formation of fine pearlite.

The transformation product at temperatures between 550 °C and 220 °C is bainite (named after E.C. Bain) consisting of a ferrite matrix in which cementite particles are embedded. Bainite formed around 450 °C appear as a feathery mass consisting of fine ferrite and cementite, which is called as upper or feathery bainite. And bainite formed in the region of around 250 °C consists of dark acicular (needle like) crystals which is called as lower or acicular bainite.

The two horizontal lines M_s and M_F at the foot of the T.T.T. diagram represent the temperatures at which the formation of martensite forms instantaneously (diffusion less) when cooled rapidly to this temperature range. As shown in the Fig. 9.4, the martensite transformation starts at a temperature 220 °C and completes at − 50 °C.

9.2.4 T.T.T. Diagram and Cooling curves

The effect of cooling rate on the formation of different reaction products such as pearlite, bainite or martensite is shown in Fig. 9.5.

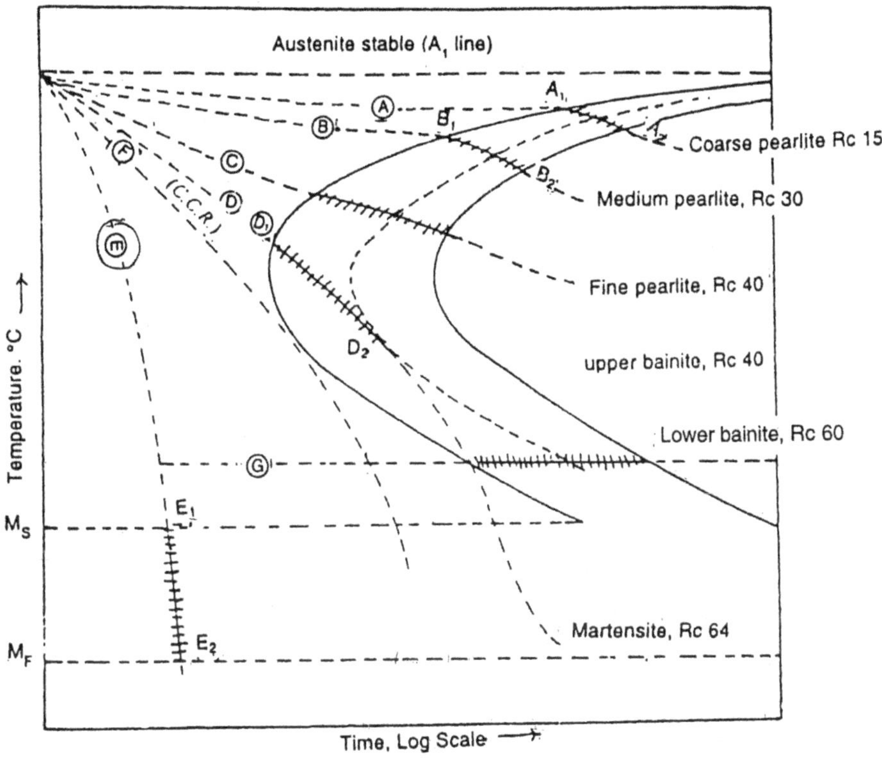

Fig. 9.5 *T.T.T. Diagram and cooling curves.*

Cooling curve A : This curve shows a very slow cooling rate typical of conventional annealing. The transformation starts at point A_1 forming coarse pearlite and ends at point A_2. The overall product will be coarse pearlite with low hardness (R/C 15)

Cooling curve B : This curve shows a faster cooling rate than curve A and may be considered as typical normalizing. The transformation starts at B_1 and ends at B_2 and the resulting product is medium pearlite. Due to faster cooling rate the pearlite formed will be finer than in case of curve A and hence overall product will be medium pearlite with a hardness of about R/C 30.

Cooling curve C : This curve is typical of slow oil quench and the resulting product consists of a mixture of medium and fine pearlite.

Cooling curve D : This curve is typical of an intermediate cooling rate and at point D_1, the austenite will start transforming into pearlite upto point D_2. And then when M_s line is crossed the remaining austenite will instantaneously transforms to martensite. Hence the final micro-structure will consist of martensite and pearlite.

Cooling curve E : This curve is typical of a drastic quench. It is cooled rapid enough to avoid transformation in the nose region. The steel remains austenitic until the M_s line is reached at point E_1. The transformation to martensite take place between point E_1 and E_2 i.e., between M_s and M_F lines. Hence the final micro structure will be entirely martensite of high hardness R/C 64.

Cooling curve F : This curve is a tangent to the nose of the T. T. T. curve. The cooling rate associated with this curve F approximates the critical cooling rate (C.C.R.). Any cooling rate slower than C.C.R. will form some softer transformation product such as pearlite or bainite. But any cooling rate equal to or faster than C.C.R. will form only martensite.

Cooling curve G : This cooling curve results in the formation of bainitic structure by cooling rapidly (represented by curve E) enough to miss the nose and then holding at a temperature at which bainite will be formed.

9.3 HEAT TREATMENT OF STEELS

The various heat treatment processes of steels are classified as follows :
- (i)   Annealing
- (ii)   Normalizing
- (iii)   Hardening
- (iv)   Tempering
- (v)   Austempering
- (vi)   Martempering
- (vii)   Case hardening
- (viii) Surface hardening

9.3.1 Annealing

Annealing operation consists of heating the steel to a certain predetermined temperature, held at this temperature for a sufficient time to allow the necessary changes to occur and then cooled slowly at a predetermined rate.

This is mainly carried to obtain the following properties.
- (i)   To relieve internal stresses induced by previous treatment such as rolling, forging, casting etc.

    (ii)   To improve machinability

    (iii)   To soften the steels

    (iv)   To refine and homogenise the structure

    (v)   To induce stable structure

    (vi)   To remove gases

    (vii)   To produce desired micro structures

    (viii)   To improve mechanical, physical, electrical and magnetie properties

Annealing processes for steels can be classified as follows :

    (i)   Stress relief annealing

    (ii)   Process annealing

    (iii)   Spheroidise annealing

    (iv)   Full annealing (or) conventional annealing.

9.3.1.1 *Stress relief annealing*

Stress relief annealing relieves or eliminates the internal or residual stresses induced by cold working, welding, casting etc.

In this process, the cold worked steel is heated to a temperature between 500 and 550 $^{\circ}$C usually below its recrystallization temperature (around 600 $^{\circ}$C), held at this temperature for about 1 to 2 hours and then cooled to room temperature in air. Due to this internal stresses are relieved without change in microstructure. If the stresses are not relieved, they might later cause warpage or even failure of the part. Stress relief annealing is best suited for both ferrous as well as non ferrous metals.

9.3.1.2 *Process annealing*

This is used in sheet and wire drawing industries to remove the effects of cold work. This heat treatment is applied after cold working to soften the steel by recrystallization and to permit further cold work.

In this process, the cold worked steel is heated to above its recrystallization temperature (600 $^{\circ}$C) i.e., below the lower critical temperature line A_1 and held at that temperature in air. Due to this, recrystallization occurs and the steel gets softened so that further cold work can be carried.

9.3.1.3 *Spheroidise annealing*

The hyper eutectoid steel consisting of pearlite and cementite net work will have poor machinability because the cutting tool can not penetrate through the hard and brittle cementite plates. Spheroidise annealing or spheroidising produces spheroidal or globular form of cementite from plates of cementite so that the machinability will be improved because the cutting tool can cut the spheroids easily.

Any of the following methods produce spheroidised structure :

(i)   Heating steel and its prolonged holding at a temperature just below the lower critical temperature line A_1 (i.e., between 650 °C and 700 °C)

(ii)   Heating and cooling alternately between temperatures that are just above and just below the lower critical temperature line A_1 –

(iii)   Heating to a temperature above the lower critical temperature line A_1 and then either cooling very slowly in the furnace or holding at a temperature just below the lower critical temperature line A_1.

9.3.1.4 *Full annealing or Conventional annealing*

The purpose of full annealing may be to refine grains, to induce softness to improve electrical and magnetic properties and in some cases, to improve machinability.

In this process, hypo eutectoid steels are heated above AC_3 temperature and hyper eutectoid steels are heated above A_1, temperature by 30 °C to 50 °C, held at this temperature for a definite period and then cooled slowly in the furnace itself at a rate of 30 °C to 200 °C per hour depending on the composition. Because of very slow cooling involved, this follows the iron-iron carbide equilibrium diagram.

Slow cooling causes the austenite to decompose at low degrees of super cooling so as to form a pearlite + ferrite structure in hypo-eutectoid steel, a pearlite + cementite structure in hyper eutectoid steel and a pearlite structure in eutectoid steel.

Hyper eutectoid steels are always annealed from above A_1 temperature and not from above Acm temperature because :

(i)   If annealed (slow cooling) from above Acm temperature, brittleness is induced in steel.

(ii)   Acm temperature is high and hence heating to above Acm results in more oxidation and decarburization of steel.

(iii)   Above Acm temperature grain coarsening of austenite occurs.

9.3.2 Normalizing

Normalizing may be employed to achieve the following :

(i)   to produce harder and stronger steel comparing to full annealing.

(ii)   to improve machinability

(iii)   to modify and refine the grain

(iv)   to homogenize the microstructure

In this process, hypo eutectoid steels are heated above the upper critical temperature AC_3 and hyper eutectoid steels are heated above Acm temperature by 30 °C to 50 °C, held at this temperature for short period and then cooled to room temperature in still air.

Air cooling, which is slightly fast as compared to furnace cooling used in case of full annealing, affects the transformation of austenite and the resultant microstructure of steels in several ways. During normalizing, there is less time for the formation of the proeutectoid constituent. Hence in hypoeutectoid steels, there will be less proeutectoid ferrite and in hyper eutectoid steels there will be less proeutectoid cementite as compared to full annealing. The faster cooling rate in normalizing also causes the formation of fine pearlite where as during full annealing coarse pearlite is formed.

Normalised carbon steel consists of pearlite and ferrite in hypoeutectoid steels and of pearlite and cementite in hypereutectoid steels.

9.3.3 Hardening

Hardening is that heat treatment process which improves the hardness of steel. This is carried to achieve the following purposes :

    (i)   to increase the hardness of steel and tool steels are hardened to improve their cutting ability.

    (ii)   to improve wear resistance of steels

    (iii)   to improve magnetic properties

In this process, the hypoeutectoid steels are heated above the upper critical temperature AC_3 and hypereutectoid steels are heated above the lower critical temperature A_1 by 30°C to 50°C, held at that temperature for sufficient time and then quenched (rapidly cooled) in water, oil or brine solution. Due to this the austenite changes instantaneously into a micro constituent called martensite (named after the scientist A. Martin who first recognised this constituent in the year 1880). Martensite is the super saturated solid solution of carbon in iron which is very hard and brittle and this imparts high hardness to steel after hardening treatment.

Hypoeutectoid steels are hardened always from above AC, temperature to convert the austenite completely to martensite. If these steels are hardened from above A_1 temperature, the soft ferrite phase that exist at this temperature will not change to martensite and it will be retained after quenching at room temperature. Ferrite being a soft phase, the hardness of steel gets reduced.

Hypereutectoid steels are hardened always from above A_1 temperature. At this temperature, along with austenite, certain amount of cementite remains in the structure.

This cementite does not change to martensite and remains in the hardened steel along with the martensite produced by austenite transformation. Since cementite being a hard phase, the hardness of hardened steel does not get reduced. But if the hardening is carried above Acm temperature, it will cause coarsening of the grain and warping of the part during quenching.

9.1.3 Quenching Medium

Quenching medium is the one into which heated metal objects are plunged in order to withdraw heat from the objects rapidly. The quenching medium must provide a cooling rate faster than the critical cooling rate (C.C.R.) to prevent austenite decomposition in the pearlite and intermediate regions. But in the martensitic transformation temperature range, cooling should be slower to avoid high internal stresses, warping of the hardened part and cracking.

The most widely used Quenching media are arranged, in order of Quenching speeds, below :

    (i)   5 to 10 percent caustic soda - very drastic quench
    (ii)   5 to 20 percent Brine (NaCl)
    (iii)   Cold water
    (iv)   Warm water
    (v)   Mineral oil
    (vi)   Animal oil
    (vii) Vegetable oil
    (viii) Air --- least drastic quench

Table 9.1 shows the relative cooling rates for some quenching media.

Sl. No.	Quenching medium	Cooling rate relative to that of water at 18°C	
		720 °C to 550 °C	**200 °C**
1.	10% Caustic soda (NaOH)	2.06	1.36
2.	Water at 0°C	1.06	1.00
3.	Water at 18°C	1.00	1.00
4.	Oil (Rapseed)	0.30	0.055
5.	Water at 100°C	0.044	0.71
6.	Air	0.028	0.007

Table 9.1 Relative cooling rates for some quenching media

Heat is extracted by quenching medium in three distinct stages of varying intensity. These stages are discussed below :

Stage A -Vapour blanket cooling stage : In the first stage, the temperature of the object is very high that the quenching medium is vaporised at the surface of the object and a thin stable vapour film or blanket surrounds the hot object. In such a condition, the cooling is by conduction and radiation through the vapour film and since vapour films are poor heat conductors, the cooling rate in this stage is relatively slow.

Stage B -Vapour transport cooling stage : This stage starts when the object has cooled to a temperature at which the vapour film is no longer stable. The vapour blanket or film is broken intermittently, allowing the liquid to touch the hot metal but soon it is pushed away from the object by vapour bubbles. The bubbles escapes from the surface and the liquid touches the hot object again. This causes violent boiling of liquid. Very rapid cooling occurs in this stage that soon brings the object surface temperature below the boiling point of the liquid.

Stage C -Liquid cooling stage : This stage starts when the surface temperature of the object reaches the boiling point of the quenching liquid. At this stage vapour no longer forms and cooling is much slower as heat is extracted by convection. Cooling rate decreases as the temperature of the metal falls. The cooling rate is slowest in this stage.

9.3.3.2 Hardenability

Hardness is a measure of resistance to plastic deformation whereas hardenability of steel is that property which determines the depth of hardening which is usually taken as the distance from the surface to the semi-martensite zone (50 percent martensite + 50 percent bainite).

The depth of hardening depends on the critical cooling rate and since critical cooling rate is not the same for the whole cross section of the object, full hardening may be achieved if the actual cooling rate even at the core exceeds the critical value.

Only the surface layer will be hardened if the actual cooling rate of this layer exceeds the critical rate and that of the core is less than the critical value. In this case the structure of the core will consist of bainite or pearlite.

The most widely used method of determining hardenability is the end quench hardenability test or the Jominy test.

9.3.3.2.1 End quench hardenability test or Jominy test

In this test, a 25.4 mm diameter by 102 mm long cylindrical bar is heated to the specified temperature. The bar is then removed from the furnace and placed on a fixture and then quenched from end in a standardised way as shown in Fig. 9.6.

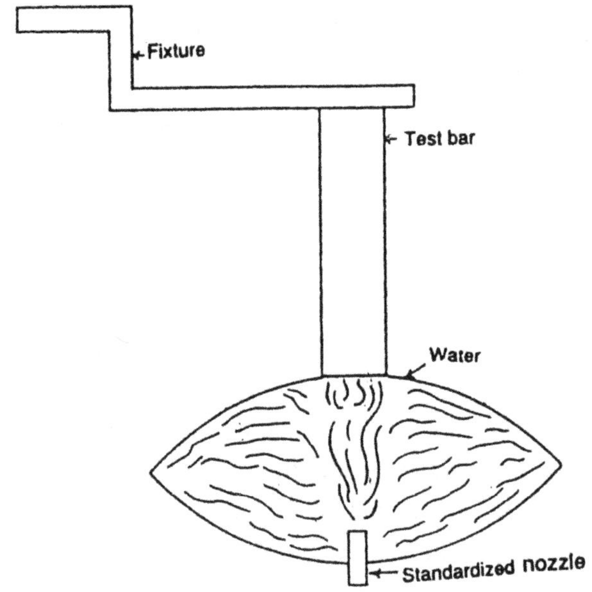

Fig. 9.6 *Jominy End Quench Test.*

The cooling rate is maximum at the quenched end of the bar where full hardening occurs and decreases gradually towards the air cooled end. After 10 minutes on the fixture, the bar is removed and two parallel flat surfaces are ground along the length of the bar. Then Rockwall C scale hardness readings are taken at 1.6 mm intervals from quenched end. As shown in Fig. 9.7, the hardness values are plotted as a function of distance from quenched end.

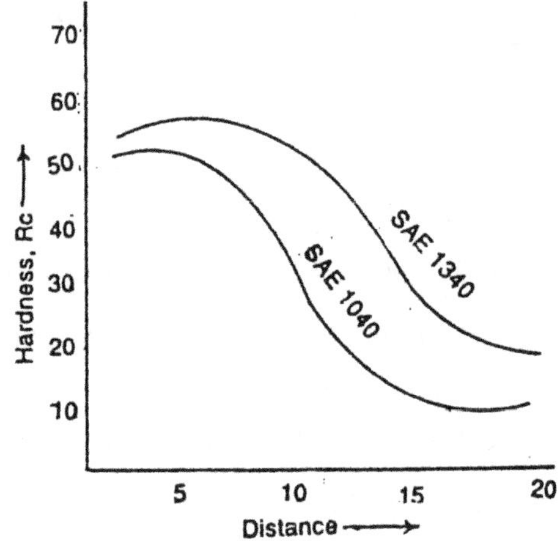

Fig. 9.7 *Jominy end quench curves for different steels.*

After plotting the curve shown in Fig. 9.7, a horizontal line may be drawn at the corresponding hardness of the semi-martensite zone for the given steel. The intersection of this horizontal line with the curve will indicate the length of the hardened zone i.e., distance from the quenched end. Using this value, the diameter of an object that will be fully hardened can he determined from a diagram shown in Fig. 9.8

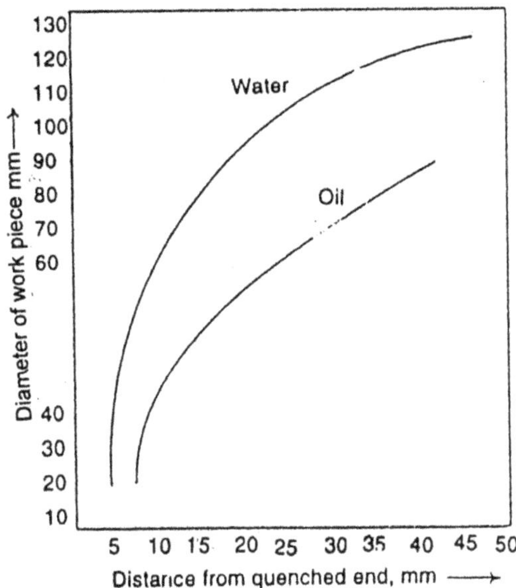

Fig. 9.8 *Determining the diameter of fully hardened objects according to the distance from quenched end.*

9.3.4 Tempering

Hardening of steel by quenching produces the microstructures consisting of martensite and retained austenite. The martensite formed during quenching is highly brittle, hard and stressed. The hardened steel is not used in this condition because cracking and distortion occurs. And also retained austenite is an unstable phase and as it changes with time, the dimensions of the object may change. Therefore, hardening is almost always followed by tempering.

Tempering consists of heating the hardened components below the lower critical temperature, held at that temperature for a specified period and then cooled to room temperature usually in air.

The purpose of tempering is :
    (i)   to relieve the internal stresses induced by quenching.
    (ii)   to reduce hardness and to increase ductility and toughness.
    (iii)   to eliminate retained austenite.

Tempering is classified into the following types :
    (i)   Low temperature tempering (100 – 200 °C)
    (ii)   Medium temperature tempering (200 – 500 °C)
    (iii)   High temperature tempering (500 – 700 °C)

9.3.4.1 *Low temperature tempering*

During low temperature tempering (100 – 200 °C), martensite decomposes and gives low carbon martensite called tempered martensite and transition iron carbide called epsilon carbide. Due to this change in micro-structure, the steel possess high strength, high hardness, low ductility, low toughness and internal stresses are relived. There is no appreciable change in the retained austenite.

9.3.4.2 *Medium temperature tempering*

During medium temperature tempering (200-500 °C) the following changes occur in microstructure :

    (i)   the epsilon carbide transforms to orthorhombic cementite.

    (ii)   the low carbon martensite (tempered martensite) transforms to ferrite.

    (iii)   the retained austenite transforms to lower bainite.

Due to these changes in microstructure, the hardness decreases and the ductility and toughness increases.

9.3.4.3 *High temperature tempering*

During high temperature tempering (500- 700 °C), the cementite particles become coarse and appear as spheroidal particles. Except this, there is no other change in the microstructure. Coarsening of cementite particles results in a slight decrease in hardness and toughness but because of spheroidal shape it possess excellent machinability.

9.3.4.4 *Temper brittleness*

Some alloy steels after being quenched and tempered in the temperature range 350-550°C and cooled slowly, loose their impact resistance and become brittle. This phenomenon is known as temper brittleness. This behaviour is due to some phase which precipitates along the grain boundaries during slow cooling. High manganese, phosphorous and chromium in steel appear to promote susceptibility to temper brittleness whereas an addition of 0.25 weight percent molybdenum will retard the effect. Temper brittleness is associated with fracture along the grain boundaries (inter-granular)

9.3.5 Austempering

Austempering is a special heat treatment process in which austenite is transformed into bainite. As shown in Fig. 9.9, in this process, steel is heated to above the austenitizing temperature and then quenched rapidly (to miss the nose of the T.T.T. diagram) in a salt bath maintained at a constant temperature above M_s point and within the bainitic range (200 to 400 °C). The steel is left in the bath until all the austenite is transformed to bainite. After complete transformation, the steel is taken out of the bath and is cooled in air or at any desired rate to room temperature. The bainite thus formed resembles tempered martensite. Hence austempered steel rarely require tempering.

Since the transformation of austenite to bainite is carried at constant temperature, it is also known as isothermal quenching or isothermal hardening. Austempered steel possess higher toughness, better ductility and hardness as compared to conventional hardening and tempering treatment. For austempering, the size of the object is restricted to relatively thin sections so that the entire object can quickly attain the temperature of the quenching bath.

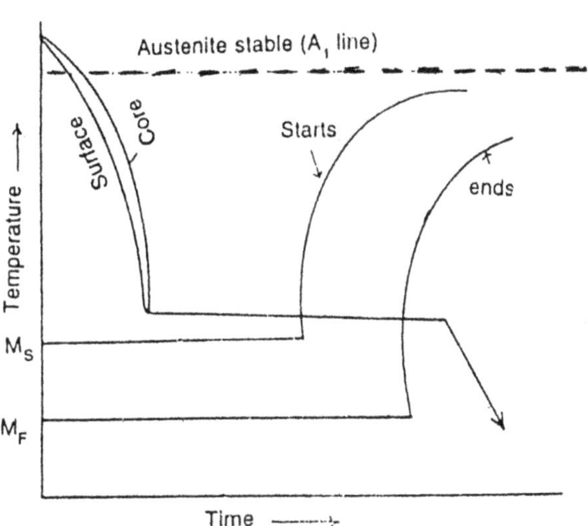

Fig. 9.9 *Austempering Treatment.*

9.3.6 Martempering

As shown in Fig. 9.10, martempering involves heating steel to above the austenitizing temperature and then quenched into a salt bath maintained at a temperature above M_s point (usually between 180 °C and 250 °C). The steel is held in the bath till the temperature throughout the section (from surface to core) becomes uniform and is equal to the bath temperature without transformation of austenite. The steel is then removed from the bath and cooled in air through the martensite range. This results in the formation of martensite with minimum residual

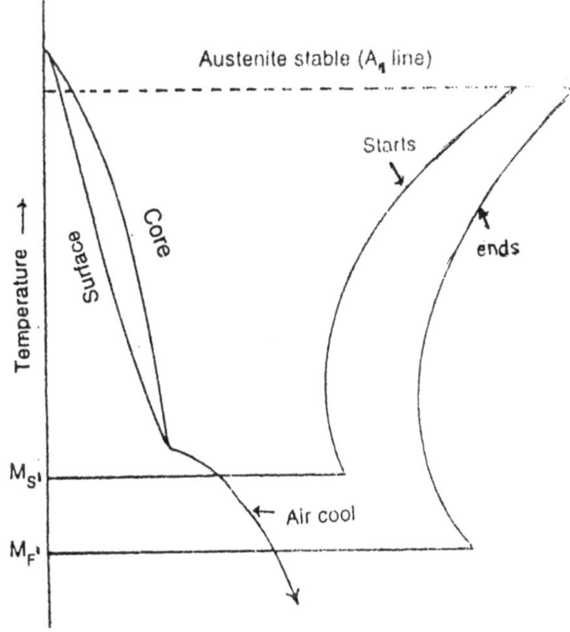

Fig. 9.10 *Martempering Treatment.*

stresses, distortion and cracking. To improve properties further, martempered steels are generally tempered.

9.3.7 Case hardening or chemical heat treatment of steels

Many industrial components such as crank pins, dies, camshafts, gears etc. require a hard wear resistant surface (called case) and soft and tough core. Both these requirements can not be met either by plain carbon steels or even by alloy steels. But however, both these properties can be achieved by employing a low carbon steel with suitable core properties and then hardening the surface layer (case) to a definite depth by adding carbon, nitrogen or both to the surface by diffusion. This treatment is known as case hardening or chemical heat treatment. This heat treatment is based on the principle that as the temperature is raised, the number of interstitial atoms and vacancies in the crystal lattice will increase, thus increasing the rat of diffusion. The total depth of the diffused layer varies with time according to the relation :

$$\text{Case depth} = K \sqrt{t}$$

where $K = \sqrt{2D}$

D is the diffusion coefficient (cm^2/s) and

t is the time of diffusion (s)

The processes generally employed for case hardening are :
- (i)   Carburising
- (ii) Nitriding
- (iii) Cyaniding
- (iv) Carbonitriding

9.3.7.1 *Carburising*

Carburising is a method of enriching the surface layer of low carbon steel with carbon in order to produce a hard case. Usually, carburising is carried out in the temperature range 900-930 °C and the surface layer is enriched with carbon upto 0.7 to 0.9 percent.

Since the solubility of carbon is more in austenitic state than in ferritic state, fully austenitic state is required for carburising. This can be achieved by heating the steel above the critical temperature. And diffusion of carbon is made by holding the heated steel in contact with carbonaceous material which may be a solid, a liquid or a gas. Depending on the form of the carburising medium (carbonaceous material), there are three general methods of carburising. These are :

    (i)   Pack carburizing

    (ii)   Liquid carburizing

    (iii) Gas carburising.

9.3.7.1.1 Pack Carburising

Pack carburising is also called as solid carburising. In this process, the machined components of low carbon steel which are to be heat treated are packed, with 80 percent charcoal and 20 percent barium carbonate ($BaCO_3$) which acts as energizer, in heat resistant boxes. The components are so placed that no component touches another or even the sides of the box. The boxes are then placed in an electric furnace and heated to a temperature between 900 - 930 °C for 6 to 8 hours. After heating the box is cooled to room temperature along with components inside the furnace. The following reactions take place at high temperature and the surface layer gets enriched with carbon

$$Ba\ CO_3 \rightarrow BaO + CO_2$$

$$CO_2 + C \rightarrow 2CO$$

$$2CO + Fe \rightarrow Fe\ (C) + CO_2$$

Case depth varies with time of heating. But generally for a carburising time of 6 to 8 hours, the case depth obtained will be 1 mm to 2 mm.

9.3.7.1.2 Liquid Carburising

Liquid carburising is carried out in molten baths, containing 20 to 50 percent sodium cyanide, 40 percent sodium carbonate and varying amounts of sodium or barium chloride. This mixture is melted and the bath temperature is maintained between 815°C and 900°C. Then the components are immersed into this molten bath for a period varying from 5 minutes to 1 hour depending on the case depth required. The following reactions cause the carbon to diffuse into the surface layer of the components.

$$BaCl_2 + 2\ NaCN \rightarrow Ba\ (CN)_2 + 2NaCl$$

$$Ba\ (CN)_2 + Fe \rightarrow Fe\ (C) + BaCN_2$$

This process gives a thin hardened layer of about 0.08 mm thick.

In this process nitrogen may also diffuse through the oxidation of sodium cyanide. The reactions are :

$$2\ NaCN + O_2 \rightarrow 2NaCNO$$

$$3NaCNO \rightarrow NaCN + Na_2CO_3 + C + 2N$$

Hence in liquid carburising, like cyaniding both carbon and nitrogen are added to the surface layer. But in cyaniding the case contains more nitrogen and less carbon and in liquid carburising the case contains more carbon and less nitrogen.

9.3.7.1.3 Gas Carburising

In this process, the components are heated to a temperature of about 900 °C for 3 to 4 hours in a gaseous atmosphere which will deposit carbon atoms by decomposition at the surfaces of the components. The gaseous atmosphere contains hydrocarbons methane (CH_4) and propane C_3H_8 diluted with a carrier gas. Carrier gases consist of a mixture of nitrogen, hydrogen and carbon monoxide.

During gas carburising the following reactions will occur and carbon gets diffused into the surface layer of components.

$$C_3H_8 \rightarrow 2CH_4 + C \text{ (Cracking of hydrocarbon)}$$

$$CH_4 + Fe \rightarrow Fe\,(C) + 2H_2$$

$$CH_4 + CO_2 \rightarrow 2CO + 2H_2$$

$$2CO + Fe \rightarrow Fe\,(C) + CO_2$$

9.3.7.1.4 Heat treatment after carburising

Since carburising involves prolonged heating at 900- 930 °C, over heating may occur which causes grain coarsening throughout the cross section. Hence specific heat treatment is required after carburising. The purpose of this heat treatment is :
   (i)   to improve the microstructure
   (ii)   to refine coarser grains of case as well as core
   (iii) to obtain high hardness at the surface
   (iv)   to break the carbide network of the case.

For fine grained steel, it is heated above AC_1 temperature and then quenched. If the steel is coarse grained then a double heat treatment may be required. In such cases the steel is first heated above AC_3 temperature and normalized to refine the grain size in the core and this also causes the dissolution of carbide network. After normalizing the steel is heated above AC_1 temperature and then quenched. The treatment causes refining of grains and hardening of case since austenite transforms to martensite at the surface. Whereas the core does not harden and remains fine grained and tough.

9.3.7.2 Nitriding

Nitriding is the process of enriching the surface of steel with nitrogen by holding for a prolonged period at a temperature of 500 °C in an atmosphere of ammonia (NH_3). Nitriding can be achieved with the components in ferritic state at 500 °C whereas in case of carburising it is required that the components must be in austenitic state.

In this process the machined and heat treated (hardening by heating to 930 °C and quenching in oil, then tempering at 650 °C to obtain the required properties in core) components are heated to a temperature of 500 °C for between 40 to 100 hours (depending on case depth) in gas tight chamber through which ammonia is allowed to circulate.

Ammonia dissociates according to the following reaction :

$$2\,NH_3 \quad \rightarrow \quad 3H_2 + 2N$$

The atomic nitrogen thus formed diffuses into iron and forms hard nitrides by combining with iron and certain alloying elements present in steel. The alloying elements having more affinity for nitrogen are aluminium, chromium and molybdenum.

The advantage of nitriding is it gives very high surface hardness and also since nitrided parts are not quenched, there is no change of distortion or cracking. The surface possess good corrosion, wear and fatigue resistance. The most disadvantage of nitriding is it takes prolonged time to achieve the required case depth and the cost of the plant is also high.

9.3.7.3 Cyaniding

During cyaniding the surface of steel is enriched with carbon and nitrogen. In this process the components are immersed in a liquid bath of 30 percent NaCN, 40 percent Na_2CO_3 and 30 percent NaCl, maintained at a temperature of 800 °C to 850 °C. Then a measured amount of air is passed through the molten bath. Sodium cyanide reacts with oxygen of the air and is oxidized. The basic reactions in the bath are :

$$2\,NaCN + O_2 \quad \rightarrow \quad 2\,NaCNO$$
$$2\,Na\,C\,NO + O_2 \quad \rightarrow \quad Na_2\,CO_3 + CO + 2N$$
$$2\,CO \quad \rightarrow \quad CO_2 + C$$

Carbon and nitrogen thus formed in atomic form diffuse into steel surface and give thin wear resistant surface layer. This process usually requires 30 to 90 minutes for completion.

After cyaniding, the components are taken out and quenched in water or oil followed by low temperature tempering.

9.3.7.4 Carbonitriding

In cyaniding the surfaces are enriched with carbon and nitrogen using liquid baths whereas in carbonitriding the surfaces are enriched with carbon and nitrogen using gaseous atmospheres.

Carbonitriding is carried out by heating the components in the temperature range of 800°C to 870 °C in a gas mixture consisting of a carburising gas and ammonia. A typical gas mixture contains about 15 percent NH_3, 5 percent CH_4 and 80 percent carrier gas. Carbon and nitrogen are diffused into the surface of the components and gives a hard wear resistant surface layer. After carbonitriding the components are quenched followed by low temperature tempering.

9.3.8 Surface Hardening

In the methods discussed above the surfaces are hardened by changing the composition of the surface with or without subsequent heat treatment. But there are other methods of surface hardening without changing the composition of the surface. These methods are :

    (i)   Flame Hardening

    (ii)   Induction Hardening

9.3.8.1 *Flame Hardening*

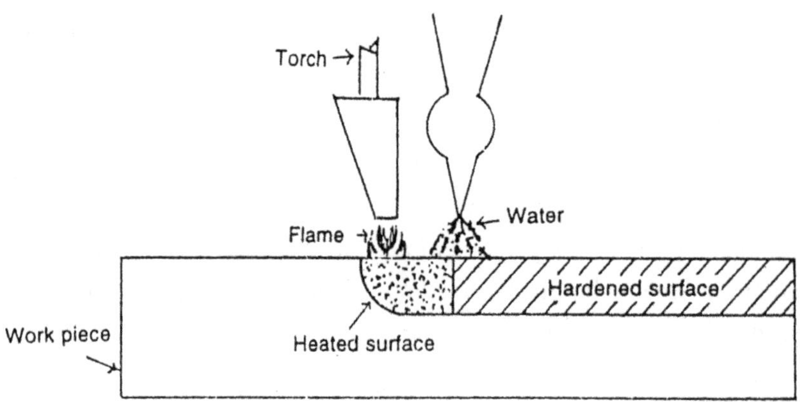

Fig. 9.11 *Principle of flame hardening.*

In this process, the heat treatable steel (usually 0.4 to 0.6 percent carbon) is rapidly heated by means of an oxyacetylene flame to above the AC_3 temperature and then quenched with water as shown in Fig. 9.11. The highly heated surfaces become hard due to the transformation of austenite to martensite whereas the core remains soft and tough. Depth of the hardened zone may be controlled by controlling the flame intensity, heating time or speed of travel of the torch. After hardening the work piece is subjected to low temperature tempering.

The disadvantages of this process is that the temperature can not be controlled accurately and over heating may cause distortion and cracking of the component.

Flame hardening can be carried in four different ways. These are :

- (i) Stationary : Here both the work piece and torch are stationary and this method is used for spot hardening of small parts.
- (ii) Progressive : Here the torch moves over stationary work piece. This method is used for large parts.
- (iii) Spinning : Here torch is stationary while the work piece rotates. This method is used for circular parts.
- (iv) Progressive Spinning: Here torch moves over a rotating work piece. Large rolls are hardened by this method.

9.3.8.2 *Induction Hardening*

The disadvantage of flame hardening i.e., over heating may be avoided by inducing heat electrically in the surface of steel.

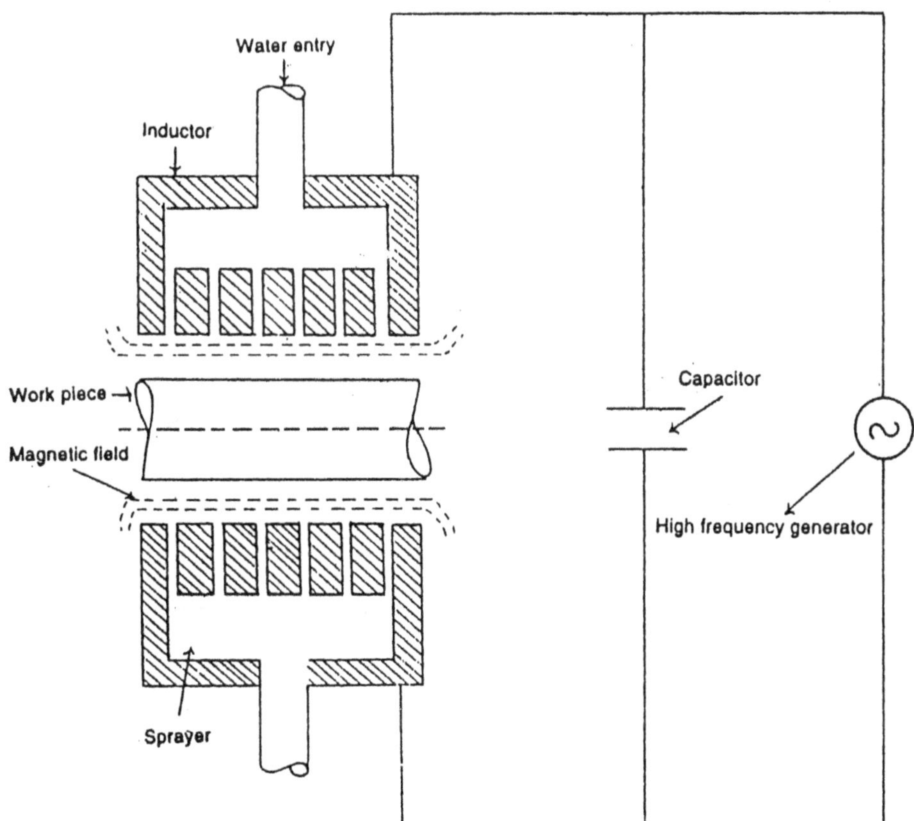

Fig. 9.12 Induction Hardening Process.

In induction hardening, the work piece (usually medium carbon steel) is heated by means of alternating magnetic field to above AC_3 temperature and then immediately quenched by a jet of cold water. Due to quenching, the austenite is transformed to martensite which produces a hard and wear resistant surface and the core remains unaltered.

In induction hardening, the heating time is only a few seconds. Heat generated in the work piece by induction is mostly confined to outer surface which is to be hardened. The depth to which heat penetrates is inversely proportional to the square root of frequency of the current. Hence the hardened depth decreases with increase in frequency of the current.

Similar to flame hardening, the induction hardened work piece is also subjected to low temperature tempering to relive stresses.

9.4 HEAT TREATMENT OF ALLOY STEELS

The heat treatment of various alloy steels is discussed in this section.

9.4.1 Heat treatment of high speed tool steel

The most widely used high speed tool steel contains 0.7 percelll carhon, 18 percent tungsten, 4 percent chromium, and 1 percent vanadium. Fig. 9.13 shows the heat treatment cycle for this grade of high speed tool steel.

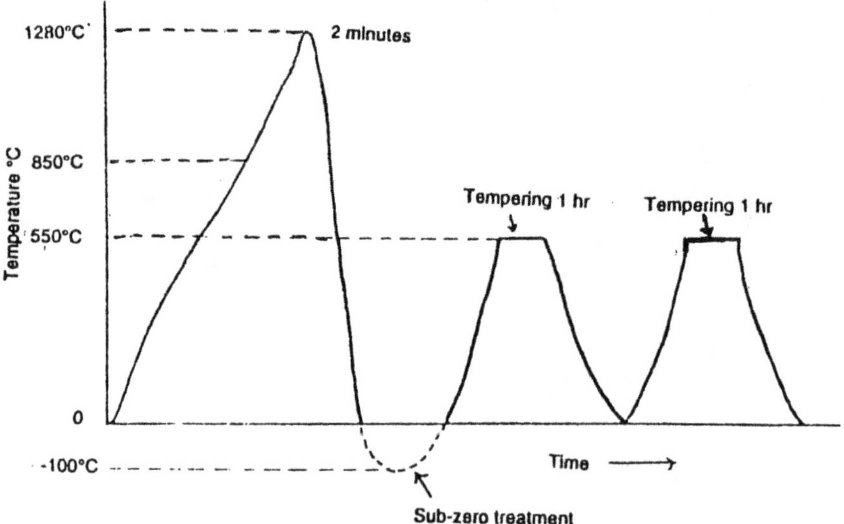

Fig. 9.13 *Hardening and Tempering of high speed tool steel.*

During hardening (as shown in Fig. 9.13) high speed tool steel is heated to a temperature between 1150 and 1350 °C. It is heated to this high temperature to dissolve more carbon and alloying elements in austenite. Since this high temperature causes oxidation and grain coarsening, heating is done in two stages. First it is preheated to 800 – 850 °C and then the steel is transferred to a salt bath furnace maintained at 1250 – 1300 °C. After heating, it is held at that temperature for 2 minutes and then quenched in oil to room temperature. The quenched steel possess the micro-structure consisting of martensite, carbides that remained undissolved during heating and retained austenite. Since the retained austenite is softer than martensite and hence lowers the cutting properties, a sub zero treatment is given after hardening to prevent stabilization of austenite (see Fig. 9.13)

As shown in Fig. 9.13. after sub zero treatment the steel is subjected to tempering. The tempering is carried out at a temperature of 550 °C. It seems multi tempering is more effective than single tempering. Multi tempering transforms the retained austenite to martensite.

9.4.2 Heat treatment of stainless steels

The following tables give the chemical composition and heat treatment temperature of a few stainless steels.

Chemical composition	Annealing temperature °C	Stress relieving temperature °C
0.08% C (max.) 18.0 -20.0% Cr 8.00 -10.0% Ni 1%Si 2.0% Mn 0.045% S 0.030% P	1010 -1125	200 -400
0.25% C 1.50% Si 24.0 – 26.0% Cr 19.0 – 22.0% Ni 2.0% Mn 0.045% S 0.030% P	1033 –1130	200 - 400

Table 9.2 *Heat treatment temperatures for a few austenitic stainless steels of different compositions*

Chemical composition	Annealing Temperature °C
0.12% C 16.0 -18.0% Cr 0.75 -1.25% MO Nb + Ta = 0.70% 1.0% Si 1.0%Mn 0.04% S 0.03% P	785 - 845
0.20% C 18.0 -23.0% Cr 1.0% Si, 1.0% Mn, 0.04% S, 0.03% P	700 -760

Table 9.3 *Heat treatment temperatures for ferritic stainless steels of different compositions.*

Chemical composition	Hardening temperature °C	Quenching medium	Tempering temperature °C
0.15% C 11.50 -13.5% Cr 1.0% Si, 1.0% Mn, 0.04% S, 0.03% P	925 -1025	Air or Oil	550 - 600 (or) 200 -350
0.15% C 12.0 -14.0% Cr 1.0% Si, 1.0% Mn, 0.04% S, 0.03% P	975 -1075	Air or Oil	200 -350

Table 9.4 *Heat treatment temperatures for martensitic stainless steels of different compositions*

9.4.3 Heat treatment of Silicon steels

Table 9.5 gives chemical compositions and heat treatment of different grades of silico - manganese spring steels.

Chemical Composition	Normalizing temperature °C	Hardening temperature °C	Quenching Medium	Tempering temperature °C
0.45 -0.55% C 1.50 -1.80% Si 0.60 -0.80% Mn 0.045% S 0.045% P	840 -870	820 -850	Water	470 -500

Table 9.5 *Heat treatment temperatures for Silico -Manganese spring steels of different compositions.*

9.4.4 Heat treatment of Manganese steels

Table 9.6 gives the chemical compositions and heat treatments of different grades of medium carbon -low manganese steels.

Chemical composition	Normalizing temperature °C	Hardening temperature °C	Quenching Medium	Tempering emperature °C
0.30 – 0.40% C, 1.70-2.00% Mn 0.15 – 0.35% Si 0.04% S, 0.04% P	850-880	830-860	Oil	530-670
0.30-0.40% C, 1.70 – 2.00% Mn 0.20 – 0.35% MO 0.15 – 0.35% Si 0.04% 5, 0.04% P	860-890	830-860	Oil	530-670
0.35 – 0.45% C, 1.70 – 2.00% Mn 0.40 – 0.60% Cr 0.08 – 0.15% V 0.15 – 0.35% Si	860-890	830-860	Oil	530-670

Table 9.6 *Heal treatment temperatures for medium carbon low manganese steels of different compositions.*

9.5 HEAT TREATMENT OF ALUMINIUM ALLOYS

The main aim of heat treating aluminium alloys is to relieve internal stresses and to increase strength and hardness. The main heat treatment processes used for aluminium alloys are :

    (i)   Annealing

    (ii)   Solution treatment

    (iii) Age hardening or Precipitation hardening

These processes are discussed below :

    (i)   Annealing : Annealing of aluminium alloys is usually carried out to relieve internal stresses induced by cold working, to increase the ductility and resistance towards shock. But the tensile and yield strength decreases with annealing.

        The aluminium alloy may be annealed by heating to a temperature of about 340 °C to 400 °C depending on the composition of the alloy, furnace cooled to a temperature of 260 °C and then withdrawn from the furnace and cooled in air to room temperature.

    (ii)   Solution treatment : In this process, the aluminium alloy is heated to a particular temperature and held at that temperature for sufficient time so that the alloying elements enter into solid solution and form single phase solid solution. After this (solutionizing), the alloy is rapidly cooled to room temperature by quenching in cold water or warm water to obtain a supersaturated solid solution.

        Consider for example aluminium -4 percent copper alloy. The solid solubility of copper in aluminium is 5.5 percent at eutectic temperature (548 °C) and decreases to 0.25 percent at room temperature. Now when this alloy is heated to about 50 °C below the eutectic temperature, all the copper in the form of $CuAl_2$ will go into solid solution and acquires a single phase structure. Rapid quenching after heating prevents the precipitation of $CuAl_2$ from the solid solution because at room temperature the alloy is not mobile enough to throw out the copper in excess of 0.25 percent and form an unstable super saturated solid solution. This results in increase of tensile strength, yield strength and hardness of the alloy.

    (iii) Age hardening or Precipitation hardening : In the solution treatment the $CuAl_2$ is forced to remain in solid solution by rapid quenching. Now with the passage of time in the cold state, the $CuAl_2$ precipitates out of the solid solution gradually. This precipitate is very fine and strengthens the alloy. This phenomenon is known as precipitation hardening or age hardening or simply ageing. Precipitation with time at room temperature is called natural ageing whereas precipitation at higher temperatures (120 °C to 200 °C) is called as artificial ageing.

The following stages are involved in the process of precipitation hardening :

(a) In the first stage the formation of zones rich in copper atoms occur. These zones or clusters are known as Guiner-Preston zones. During this process the mechanical properties are improved.

(b) Formation of transition structures in the form of modified Guiner-Preston zones (GP - II Zones) and intermediate phases. This may give rise to maximum strengthening in the alloy.

(c) formation of stable phases from transition phases occur.

9.6 HEAT TREATMENT OF COPPER

Only annealing treatment is given to pure copper. The purpose of annealing is to achieve the original ductility and softness in cold worked copper. In annealing of pure copper, it is heated to about 600 °C, held at that temperature for sufficient period and then quenched in cold water. Water quenching removes scale formation and gives clean surface of copper.

9.6.1 Heat treatment of brasses

Brasses are also subjected to only annealing treatment after cold working. In annealing of brass, it is heated to a temperature between 650 °C and 700 °C, held at that temperature for some time and cooled to room temperature at any suitable rate. This causes the cold worked brass to regain the original ductility and toughness.

Brasses can also be annealed at low temperature (300 °C) for about one hour to remove internal stresses and hence reduce the tendency for season cracking.

9.7 MODEL QUESTIONS

1. Define heat treatment process and explain the purpose of heat treatment.

2. What is T.T.T. diagram and explain the importance of T.T.T. diagram?

3. Explain the various steps involved in constructing a T.T.T. diagram.

4. What is Reaction curve? How is it useful in the construction of T.T.T. diagram.

5. Draw T.T.T. diagram for an eutectoid steel and explain the decomposition of austenite into pearlite, bainite and martensite.

6. Represent heat treatment processes on T. T. T. diagram with the help of cooling curves.

7. List out the various heat treatment processes of steel.

8. Distinguish between annealing and normalizing.

9. What is the purpose of annealing? How annealing processes are classified? Explain them.

10. Distinguish between process annealing and spheroidise annealing.

11. Explain full annealing process and normalizing.

12. What is the purpose of hardening steel?

13. What is quenching medium? Describe various industrial quenching mediums.

14. Explain the three distinct stages of heat extraction by quenching medium.

15. Define hardenability ? How hardenability is determined ?

16. Explain Jominy test or end quench hardenability test.

17. What is tempering and what is the purpose of tempering steel ?

18. Explain in brief temper brittleness.

19. Distinguish between Austempering and Martempering

20. What is case hardening? Explain the importance of case hardening.

21. Distinguish between carburising and nitriding.

22. Differentiate between pack carburising, liquid carburising and gas carburising.

23. Distinguish between Cyaniding and carbonitriding.

24. Explain the following :
    (i)   Flame hardening
    (ii) Induction hardening

25. Explain the heat treatment cycle with neat sketch for high speed tool steel.

26. Explain the following with respect to heat treatment of aluminium :
    (i)   Solution treatment
    (ii) Age hardening (or) precipitation hardening.

INDEX

www.ingramcontent.com/pod-product-compliance
Lightning Source LLC
LaVergne TN
LVHW080848240726

843527LV00052B/275